Lentils

Potential Resources for Enhancing Genetic Ga

Lentils

Potential Resources for Enhancing Genetic Gains

Edited by

Mohar Singh

ELSEVIER

ACADEMIC PRESS

An imprint of Elsevier

Academic Press is an imprint of Elsevier
125 London Wall, London EC2Y 5AS, United Kingdom
525 B Street, Suite 1650, San Diego, CA 92101, United States
50 Hampshire Street, 5th Floor, Cambridge, MA 02139, United States
The Boulevard, Langford Lane, Kidlington, Oxford OX5 1GB, United Kingdom

Library of Congress Cataloging-in-Publication Data
A catalog record for this book is available from the Library of Congress

British Library Cataloguing-in-Publication Data
A catalogue record for this book is available from the British Library

ISBN 978-0-12-813522-8

For information on all Academic Press publications
visit our website at https://www.elsevier.com/books-and-journals

Publisher: Andre Gerhard Wolff
Acquisition Editor: Nancy Maragioglio
Editorial Project Manager: Ruby Smith
Production Project Manager: Divya Krishna Kumar
Cover Designer: Mark Rogers

Typeset by SPi Global, India

Contents

Contributors xi
About the Editor xv
Preface xvii
About the Book xix

1. Introduction
Mohar Singh

2. Origin, Distribution, and Gene Pools
Aqeel Hasan Rizvi, Muraleedhar Aski, Ashutosh Sarker, Harsh Kumar Dikshit, Prachi Yadav

2.1 Introduction 7
2.2 History 8
2.3 Origin 9
2.4 Taxonomy, Morphology and Floral Biology 11
2.5 Distribution 13
2.6 Gene Pool 14
2.7 Domestication 16
References 17

3. Genetic Resources: Collection, Conservation, Characterization and Maintenance
Nikhil Malhotra, Sweety Panatu, Badal Singh, Narender Negi, Dayal Singh, Mohar Singh, Rahul Chandora

3.1 Introduction 21
3.2 Species Distribution and Gene Pools 22
3.3 Germplasm Conservation 22
3.3.1 In-Situ Conservation 23
3.3.2 Ex-Situ Conservation 24
3.3.3 Duplicates 30
3.3.4 Safety Duplicate 30
3.4 Germplasm Characterization and Evaluation 30
3.5 Development of Core Collections 34
3.6 Conclusion and Future Prospects 35
References 36
Further Reading 41

4. Conventional Genetic Manipulations

Rakesh Kumar Chahota, Tilak Raj Sharma, Shyam Kumar Sharma

4.1 Introduction 43
4.2 Taxonomy 44
4.3 Manipulation in Lentils 46
4.3.1 Cytogenetic Manipulation 46
4.3.2 Genetic Manipulations 47
4.4 Breeding Strategy 51
4.5 Conclusions 52
References 52
Further Reading 55

5. Embryo Rescue and Cytogenetic Manipulation

Dorin Gupta, Rebecca Ford, Prabhakaran Sambasivam, Sajitha Biju

5.1 Introduction 57
5.2 Taxonomy of the *Lens* Genus 58
5.3 Cytogenetics of the *Lens* Genus 59
5.3.1 Cytogenetic Manipulations and Ploidy Status 61
5.3.2 Chromosome Banding Techniques 62
5.3.3 Cytogenetic and Mutation Studies 63
5.3.4 Molecular Cytogenetics 64
5.4 Wide Hybridization, Embryo-Ovule Rescue and Haploid Production 67
5.4.1 Embryo-Ovule Rescue 67
5.4.2 Haploid Breeding 73
5.5 Conclusions 74
References 74
Further Reading 81

6. Gene Pyramiding and Multiple Character Breeding

Maneet Rana, Ankita Sood, Waseem Hussain, Rahul Kaldate, Tilak Raj Sharma, R.K. Gill, Shiv Kumar, Sarvjeet Singh

6.1 Introduction 83
6.2 Conventional Breeding Approaches to Integrate Multiple Traits 85
6.2.1 Important Traits to be Focused Upon 85
6.2.2 Knowledge of Genetic Association Between Target Traits 88
6.2.3 Breeding Schemes for Multiple Traits 89
6.3 Molecular Breeding Approaches to Introgress Multiple Target Genes Governing Same or Different Traits 97
6.3.1 Molecular Marker Identification, Linkage Map Development and QTL Mapping 99
6.3.2 Genetic Linkage Maps of *Lens* 100

6.3.3 QTL Mapping 100
6.3.4 Marker-Assisted Backcrossing 103
6.3.5 Gene Pyramiding or Gene Stacking 109
6.3.6 Gene Pyramiding Through Multiple Parents Crossing 109
6.3.7 Marker-Assisted Backcrossing-Gene Pyramiding 110
6.3.8 Marker-Assisted Recurrent Selection 112
6.3.9 Genomic Selection to Improve Multiple Traits 114
6.4 Conclusions 116
References 116
Further Reading 124

7. **Molecular Marker Assisted Gene Pyramiding**

Jitendra Kumar, Shiv Kumar, Debjyoti Sen Gupta, Sonali Dubey, Sunanda Gupta, Priyanka Gupta

7.1 Introduction 125
7.2 Molecular Markers 126
7.3 Trait-Specific Mapping Populations 127
7.4 Development of Molecular Linkage Maps 128
7.5 Marker-Trait Linkage Analysis (Mapping of Genes/QTL) Using Biparental Populations 131
7.6 Marker-Trait Association Analysis Using Natural Diverse Population 131
7.7 Present Status and Future Prospects of Marker-Assisted Selection 134
7.7.1 Marker-Assisted Evaluation of Breeding Materials 134
7.7.2 Marker-Assisted Backcrossing Program 134
References 135
Further Reading 139

8. **Genetic Transformation**

Rakha Hari Sarker, Subroto Kumar Das, Kishwar Jahan Shethi, M. Imdadul Hoque

8.1 Introduction 141
8.1.1 Lentil—A Popular and Important Pulse Crop 141
8.1.2 Constraints to Lentil Production—Biotic and Abiotic Stress Factors 142
8.1.3 Improvement of Lentils Through Conventional Breeding 144
8.1.4 Application of Biotechnology for Improvement of Lentil 145
8.2 Protocols for In-Vitro Regeneration of Plantlets 150
8.2.1 Plant Materials Used for In-Vitro Regeneration 151
8.2.2 Preparation of Explants for In-Vitro Regeneration 151
8.2.3 Development of In-Vitro Shoots 151
8.2.4 Development of In-Vitro Roots From Excised Regenerated Shoots 151

8.3 Protocol for In-Vitro Flower and Pod Formation 152
8.3.1 Pollen Viability and Pollen Tube Growth Following the Self-Pollination 152
8.3.2 Propagation of In-Vitro Grown Seeds 153
8.4 Protocol for *Agrobacterium*-Mediated Genetic Transformation 153
8.4.1 Gene Construct I (pBI121*GUS npt*II) 153
8.4.2 Gene Construct II (pGIIMHvst N-chitin) 155
8.4.3 Gene Construct III (pCAMBIA2300enh35SAFP) 155
8.4.4 Steps of Transformation and Selection of Transformants 155
8.4.5 GUS Histochemical Assay 156
8.4.6 Molecular Analysis in Detecting Expression of Transgenes 156
8.4.7 Analysis of Transgenic Progenies 157
8.5 In-Vitro Regeneration of Shoots 158
8.6 In-Vitro Regeneration of Roots 160
8.7 In-Vitro Flowering and Pod Formation in Lentils 165
8.7.1 Development of In-Vitro Flowers 166
8.7.2 In-Vitro Pod and Seed Development 170
8.7.3 In-Vitro Germination of Seeds and Establishment of Plantlets Into Soil 172
8.8 *Agrobacterium*-Mediated Genetic Transformation Using Marker Genes 175
8.8.1 Responses of Various Explants of Lentils Toward *Agrobacterium*-Mediated Genetic Transformation 177
8.8.2 Regeneration and Selection of Transformed Shoots 178
8.8.3 In-Vitro Flowering and Pod Formation Using Transformed Shoots 184
8.8.4 Analysis of Transgene Integration 188
8.9 Genetic Transformation Using Antifungal Gene Constructs 188
8.9.1 Transformation With EHA105/pGII-pSOUPvst N-chitin 189
8.9.2 Transformation With LBA4404/pCAMBIA2300 enh35S AFP 191
8.10 Conclusions 195
Acknowledgements 195
References 195

9. The Lentil Economy in India

Nisha Varghese, Atul Dogra, Ashutosh Sarker, Aden Aw Hassan

9.1 Introduction 203
9.2 Global Pulse Economy: An Overview 204
9.3 Lentils: An Overview 206
9.4 Indian Pulse Scenario 209
9.4.1 Trade Policy on Pulses 210
9.4.2 SWOT Analysis of Domestic Production Vis-à-Vis Import of Pulses 211
9.4.3 The Dal Shock of 2015–16 211

9.5 Lentil Production in India 211
9.5.1 Economic Classification 211
9.5.2 The Lentil Trade Scenario 213
9.6 Comparative Economics of Cost of Production of Lentils in Madhya Pradesh 213
9.7 Comparative Economics of Lentils and Competing Crops 215
9.7.1 Prices of Lentils in India 215
9.7.2 Opportunities in Rice Fallows 217
9.8 Conclusions 218
References 218
Further Reading 219

Index 221

Contributors

Numbers in parentheses indicate the pages on which the authors' contributions begin.

Muraleedhar Aski (7), Indian Agricultural Research Institute, Division of Genetics, New Delhi, India

Sajitha Biju (57), School of Food and Agriculture, Faculty of Veterinary and Agricultural Sciences, University of Melbourne, Dookie, VIC, Australia

Rakesh Kumar Chahota (43), Department of Agricultural Biotechnology, CSK Himachal Pradesh Agriculture University, Palampur, India

Rahul Chandora (21), ICAR-National Bureau of Plant Genetic Resources, Regional Station, Shimla, India

Subroto Kumar Das (141), Department of Botany, University of Barisal, Bangladesh

Harsh Kumar Dikshit (7), Indian Agricultural Research Institute, Division of Genetics, New Delhi, India

Atul Dogra (203), International Center for Agricultural Research in the Dry Areas, South Asia & China Regional Program, New Delhi, India

Sonali Dubey (125), Division of Crop Improvement, Indian Institute of Pulses Research, Kanpur, India

Rebecca Ford (57), Environmental Futures Research Institute, School of Natural Sciences, Griffith University, Nathan, QLD, Australia

R.K. Gill (83), Department of Plant Breeding and Genetics, Punjab Agricultural University, Ludhiana, India

Dorin Gupta (57), School of Food and Agriculture, Faculty of Veterinary and Agricultural Sciences, University of Melbourne, Dookie, VIC, Australia

Debjyoti Sen Gupta (125), Division of Crop Improvement, Indian Institute of Pulses Research, Kanpur, India

Sunanda Gupta (125), Division of Crop Improvement, Indian Institute of Pulses Research, Kanpur, India

Priyanka Gupta (125), Division of Crop Improvement, Indian Institute of Pulses Research, Kanpur, India

Aden Aw Hassan (203), International Center for Agricultural Research in the Dry Area, ICARDA, Amman, Jordan

M. Imdadul Hoque (141), Plant Breeding and Biotechnology Laboratory, Department of Botany, University of Dhaka, Dhaka, Bangladesh

Waseem Hussain (83), Department of Agronomy and Horticulture, University of Nebraska, Lincoln, NE, United States

Rahul Kaldate (83), Department of Agricultural Biotechnology, CSKHPKV, Palampur, India

Jitendra Kumar (125), Division of Crop Improvement, Indian Institute of Pulses Research, Kanpur, India

Shiv Kumar (83, 125), Biodiversity and Integrated Gene Management Program, International Center for Agricultural Research in the Dry Areas (ICARDA), Rabat, Morocco

Nikhil Malhotra (21), ICAR-National Bureau of Plant Genetic Resources, Regional Station, Shimla, India

Narender Negi (21), ICAR-National Bureau of Plant Genetic Resources, Regional Station, Shimla, India

Sweety Panatu (21), ICAR-National Bureau of Plant Genetic Resources, Regional Station, Shimla, India

Maneet Rana (83), Division of Crop Improvement, ICAR-Indian Grassland and Fodder Research Institute, Jhansi, India

Aqeel Hasan Rizvi (7), International Center for Agricultural Research in the Dry Areas, South Asia & China Regional Program, New Delhi, India

Prabhakaran Sambasivam (57), Environmental Futures Research Institute, School of Natural Sciences, Griffith University, Nathan, QLD, Australia

Ashutosh Sarker (7, 203), International Center for Agricultural Research in the Dry Areas, South Asia & China Regional Program, New Delhi, India

Rakha Hari Sarker (141), Plant Breeding and Biotechnology Laboratory, Department of Botany, University of Dhaka, Dhaka, Bangladesh

Tilak Raj Sharma (43, 83), ICAR-Indian Institute of Agricultural Biotechnology, Ranchi, India

Shyam Kumar Sharma (43), CSIR-Institutes of Himalayan Bioresources and Technology, Palampur, India

Kishwar Jahan Shethi (141), Plant Breeding and Biotechnology Laboratory, Department of Botany, University of Dhaka, Dhaka, Bangladesh

Badal Singh (21), ICAR-National Bureau of Plant Genetic Resources, Regional Station, Shimla, India

Dayal Singh (21), ICAR-National Bureau of Plant Genetic Resources, Regional Station, Shimla, India

Mohar Singh (1, 21), ICAR-National Bureau of Plant Genetic Resources, Regional Station, Shimla, India

Sarvjeet Singh (83), Department of Plant Breeding and Genetics, Punjab Agricultural University, Ludhiana, India

Ankita Sood (83), Department of Plant Breeding and Genetics, Punjab Agricultural University, Ludhiana, India

Nisha Varghese (203), School of Extension and Development Studies, Indira Gandhi National Open University, New Delhi, India

Prachi Yadav (7), Indian Agricultural Research Institute, Division of Genetics, New Delhi, India

About the Editor

Dr. Mohar Singh is currently working as principal scientist at the National Bureau of Plant Genetic Resources Regional Station in Shimla, India. He obtained his doctoral degree in plant breeding from the Himachal Pradesh Agricultural University in Palampur, India. He has been working on prebreeding and genetic enhancement in lentils and chickpeas for the last several years. He has identified useful gene sources for various traits of interest in wild lentils and chickpeas, and some of them have been introgressed into the cultivated backgrounds for diversification of the cultivated gene pool. He has published more than 60 research articles in reputed international journals, namely, *Nature Scientific Reports, Plant Breeding, Crop Science, Euphytica, PLoS ONE, DNA Research, Frontiers in Plant Science, Journal of Genetics, Journal of Genetics and Breeding, Plant Genetic Resources: Characterization and Utilization*, and *Advances in Horticulture Science*. He also holds two textbooks and four edited books to his credit.

Mohar Singh

Preface

Most cultivated legume crop species, including lentils, have been exploited to their maximum level of productivity. It is projected that by 2050, the world's population will be more than 9 billion, requiring an astonishing increase in food production. To attain further breakthroughs in enhancing yield and improving stability of future crop varieties, new sources of genes and alleles must be identified in plant genetic resources (PGR), including crop wild relatives (CWRs), and incorporated into elite genetic backgrounds. This will permit new plant types to be tailored for different conditions. In the face of climate change, this great resource may prove to be instrumental for future food and nutritional security. Therefore, it is imperative to motivate crop breeders to search for new sources of variation in untapped gene pools, and to identify target traits of interest using appropriate tools and techniques in order to make the selection more efficient and reliable. In this context, we have gathered the rather scattered research in this useful area in the form of an edited collection, a compilation that should be of great value to the lentil researchers across the world. The book consists of a total of nine chapters contributed by eminent researchers from various highly reputed institutions of the world. An introductory chapter describes some key issues linked to bottlenecks and the potential impact of species utilization on current trends of interspecific hybridization. Subsequent chapters deal with different aspects related to widening the genetic base of the elite gene pool for enhancing genetic gains of lentil cultivars. The editor is extremely thankful to all authors for their significant contributions to this volume. The entire process of preparing the manuscript was marked by cordial collegiality. I am also indebted to the staff of Academic Press for their excellent professional support in the completion of this project. Despite several rounds of proofreading and our best efforts, the book may still have some scientific, technical, or printing errors. I would appreciate it if these omissions were brought to my notice, so that they may be addressed in future editions.

Mohar Singh
Editor

About the Book

The multiauthor, edited book entitled, *Lentils: Potential Resources for Enhancing Genetic Gains*, consists of comprehensive coverage of important topics including origin, distribution, and gene pools; genetic resources: collection, conservation, characterization and maintenance; conventional cytogenetic manipulation; embryo rescue and cytogenetic manipulations; gene pyramiding and multiple-character breeding; molecular markers-assisted gene pyramiding; genetic transformation, and the lentil economy in India. All chapters were written by eminent lentil researchers from prominent institutions, and they provide a rare insight into the crop-specific constraints and prospects, drawing from their rich experience. An edited book, therefore, will be a useful source of information to lentil scientists, teachers, students, policy planners, and developmental experts alike.

Chapter 1

Introduction

Mohar Singh
ICAR-National Bureau of Plant Genetic Resources, Regional Station, Shimla, India

Domesticated lentils (*Lens culinaris* ssp. *culinaris*) is an annual, herbaceous, self-pollinating, true diploid ($2n = 2\times = 14$) species with an estimated genome size of 4063 Mbp/C (Arumuganathan and Earle, 1991). The crop is one of the first domesticated grain legume species originated from the Near East center of origin (Zohary, 1999) and it is the most appreciated grain legume of the Old World (Smartt, 1990). It is an important cool-season legume species grown in Mediterranean and semi-arid climates. It provides an affordable source of dietary protein (22%–25%), minerals (K, P, Fe, and Zn), carbohydrates, and vitamins for human nutrition (Bhatty, 1988; Kumar et al., 2018). Lentil grains are also rich in lysine and tryptophan content (Erskine et al., 1990). The genus *Lens* belongs to family *Fabaceae*, and a total of seven annual species have been recognized, including the cultivated species *L. culinaris* subsp. *culinaris*. The other wild taxa include *L. culinaris* ssp. *odemensis* Ladizinsky; *L. culinaris* ssp. *orientalis* (Boiss) Ponert; *L. ervoides* (Brign) Granade; *L. lamottei* Czefr; *L. nigricans* (Bieb) Godron; and *L. tomentosus* Ladizinsky. As far as the cross-compatibility of *Lens* taxa is concerned, *L. culinaris* subsp. *orientalis* is fully cross-compatible with the domesticated lentil (Roberson and Erskine, 1997) and has been proposed as its putative ancestor (Barulina, 1930; Mayer and Soltis, 1994). Globally, the lentil ranks sixth in production among grain legumes after dry beans, peas, chickpeas, faba beans, and cowpeas (FAO, 2015). However, the world's lentil production constituted 6% of total dry pulse production during 2010–15, with an average productivity of 926 kg/ha. India is the biggest lentil-producing country in the world, followed by Canada and Turkey, which collectively accounted for 66% of total global lentil production (FAOSTAT, 2016). Further, average lentil yield in Asia is 817 kg/ha, which is significantly below the world average of 926 kg/ha. Lentils, despite their tremendous significance in human food, animal feed, and cropping systems (in the Indian sub-continent, West Asia, Ethiopia, North Africa, parts of Southern Europe, Oceania, and North America), have remained an under-exploited and under-researched crop until recently. Modern lentil commercial varieties have some superiority over traditional ones in terms of their yield potential and disease-resistance. A small

Lentils. https://doi.org/10.1016/B978-0-12-813522-8.00001-7

number of improved local landraces have contributed significantly to the development of a majority of improved varieties through pure lines and mass selection, following hybridization between lines adapted to specific environmental conditions across lentil-growing regions. Notwithstanding the number of lentil varieties released, there has been slow progress in production and productivity of this important crop over the decades. Besides a very high influence of environmental factors, cultivated lentil germplasm has low genetic variation as compared to wild species (Ford et al., 1997; Duran et al., 2004). Among different accounts across lentil-growing regions in India, the pedigree analysis of 35 released lentil varieties has been traced back to only 22 ancestors, and the top 10 contributed 30% to the genetic base of released cultivars (Kumar et al., 2003). This situation could lead to crop vulnerability to pests, disease epidemics, and unpredictable climatic factors, limiting progress in increasing lentil production. Furthermore, the narrow genetic base of lentil varieties owes their vulnerability to several biotic and abiotic stresses and loss of yield-contributing traits because of their cultivation on marginal lands, mostly in developing countries (including India). To meet the dietary requirements of a growing human population, consolidated efforts are necessary to increase the genetic potential of existing lentil varieties. Therefore, there is an immediate need to broaden the genetic base of lentil cultivars by introgression of diverse gene sources, which are available in distantly related wild *Lens* taxa. To broaden the genetic base and maximize gains from selection, it is imperative to accumulate favorable genes and alleles from the potential germplasm in elite lentil backgrounds.

To enrich the genetic resource base, it is clear that germplasm must be collected from the diversity rich regions across the globe. The establishment of ex-situ germplasm collections (gene bank) has been the result of rigorous global efforts over several decades to conserve diverse genetic resources, including crop wild relatives (CWRs). In the case of lentils, the International Centre for Agricultural Research in Dry Areas (ICARDA) holds the largest collection of lentil germplasm, including wild species collected from more than 46 countries (Furman et al., 2009). The first ex-situ gene bank of lentil collection including wild relatives was established in the Vavilov Institute for plant industry in Russia. The survey of global ex-situ collection of lentil germplasm accessions lists more than 43,214 and about 40,000 other accessions, which are preserved in the different gene banks around the world (GCDT, 2016). For wild lentil germplasm collection, ICARDA has undertaken more than 120 explorations and collected more than 400 wild accessions from diversity-rich areas of the world (Redden et al., 2007). Turkey has the largest holdings of lentil landraces (Atikyilmaz, 2010), followed by Nepal and Pakistan (Sultana and Ghafoor, 2008), Bangladesh, Spain, Syria, Ethiopia, and China (Fikiru et al., 2007; Liu et al., 2008). An exact status of on-farm in-situ conservation of diverse germplasm is not well documented (Furman et al., 2009). Recently, wild species of lentils have received positive attention for being an invaluable genetic resource for widening the genetic base of cultivated lentil varieties. Because these species hold a wealth of useful genes and alleles, it is possible to break yield barriers and enhance tolerance to various biotic and abiotic stresses. The important regions for in-situ on-farm conservation of wild germplasm comprise West Turkey for *L. nigricans*, Southeast Turkey, Southwest Syria, and Jordan for *L. culinaris* ssp. *orientalis*, South Syria for *L. culinaris* ssp. *odemensis*, and the border areas of Turkey and Syria for *L. ervoides* (Ferguson et al., 1998).

Each of these species has its own morphological characteristics and shows specific ecological affinities and typical geographic distribution. *L. culinaris* ssp. *orientalis* and *odemensis* are members of the primary gene pool, whereas *L. tomentosus*, *L. lamottei*, *L. nigricans*, and *L. ervoides* species belong to the secondary gene pool. A recent study using Genotyping by Sequencing (GBS) method based on the phylogenetic tree and STRUCTURE analysis identified four gene pools, namely, *L. culinaris*/*L. orientalis*/*L. tomentosus*, *L. lamottei*/*L. odemensis*, *L. ervoides*, and *L. nigricans*, which form primary, secondary, tertiary, and quaternary gene pools, respectively. Germplasm enhancement approaches have also been used to study variation among important traits of interests, namely, high pods/plant, seeds/pod, number of clusters/plant, harvest index (%), biological yield (g), and early maturity, etc. Sincere efforts have also been made to characterize and evaluate various agro-morphological traits in large numbers of lentil accessions, including global wild accessions, for which remarkable variation was recorded. Many attempts have been made in the past to transfer desirable genes from wild lentil species to cultigens using embryo rescue technique and cold-tolerant genes from *L. orientalis* and some disease-resistant genes from *Lens ervoides* have also been transferred to the cultivated lentil. However, an understanding of the genetic potential of both cultivated and wild lentils is essential for mining useful genetic variations, as well as for adding considerable resources to lentil breeding programs. The chapter also addresses the knowledge on available lentil genomics derived through traditional cytogenetics and the potential to improve our understanding through applications of modern cytogenetic manipulations, including the use of fluorescence in-situ hybridization (FISH), genome in-situ hybridization (GISH), and distant hybridization, coupled with embryo ovule rescue and haploid breeding to widen the genetic resources of lentil.

Furthermore, molecular breeding technologies including marker-assisted selection, marker-assisted backcrossing, marker-assisted recurrent selection, gene pyramiding, marker-assisted backcross gene pyramiding, and genomic selection have been used to introgress single or multiple genes. Multiple-trait selection using selection indices based on information from both phenotypes and markers distributed across the whole genome has recently been practiced in various crops including lentils. Multiple-trait selection is a realistic approach that can be exploited in lentil breeding programs to improve multiple traits simultaneously. There has been considerable progress on in-vitro regeneration as well as *Agrobacterium*-mediated genetic transformation to integrate fungal diseases resistant gene in *microsperma* varieties of lentils. A complete protocol for *Agrobacterium*-mediated genetic transformation was established by utilizing *GUS* and *npt*II genes. Among various explants studied, cotyledon-attached decapitated embryo (CADE) appeared to have the best response toward in-vitro regeneration compatible to *Agrobacterium*-mediated genetic transformation. Those in-vitro regenerated shoots failed to produce effective roots were subjected to induce in-vitro flowers as well as seeds to develop an alternative regeneration system for lentil avoiding the in-vitro root formation stage. For enhancing the utilization of lentil germplasm, a Focused Identification of Germplasm Strategy (FIGS) has also been developed. In this strategy, germplasm for a target trait is identified from that region, where it is frequently occurred. The strategy has been useful in identification of genotypes having desirable traits such as tolerance to biotic and abiotic stresses, superior grain quality and nutritional traits, improved grain size, and early maturity. Thus, the development of core sets of diverse origin after their screening at multiple locations for useful traits may lead to their utilization in the improvement of yield and productivity.

REFERENCES

Arumuganathan, K., Earle, E.D., 1991. Nuclear DNA content of some important plant species. Plant Mol. Biol. Rep. 9, 208–218.

Atikyilmaz, N., 2010. In: Grain legumes genetic resources activities in Turkey. ECPGR Working Group on Grain Legumes, 4th Meeting, 16–17 November 2007, Lisbon. Available from: http://www.ecpgr.cgiar.org/workgroups/grain-legumes/CReps/Turkey-report.pdf.

Barulina, H., 1930. Lentil of the U.S.S.R. and of other countries. Bull. Appl. Bot. Plant Breed. Suppl. 40, 319.

Bhatty, R.S., 1988. Composition and quality of lentil (*Lens culinaris* Medik.): a review. Can. Inst. Food Sci. Technol. J. 21, 144–160.

Duran, Y., Fratini, R., Garcia, P., Perez de la Vega, M., 2004. An inter-sub-specific genetic map of *Lens*. Theor. Appl. Genet. 108, 1265–1273.

Erskine, W., Rihawe, S., Capper, B.S., 1990. Variation in lentil straw quality. Anim. Food Sci. Technol. 28, 61–69.

FAO, 2015. FAOSTAT: Food and Agriculture Organization of the United Nations, Rome.

FAOSTAT, 2016. FAO: Food Price Indices, May 2011. Available from: http://www.fao.org.

Ferguson, M.E., Ford-Lloyd, B., Robertson, L.D., Maxted, N., Newbury, H.J., 1998. Mapping the geographical distribution of genetic variation in the genus *Lens* for the enhanced conservation of plant genetic diversity. Mol. Ecol. 7, 1743–1755.

Fikiru, E., Tesfaye, K., Bekele, E., 2007. Genetic diversity and population structure of Ethiopian lentil (*Lens culinaris* Medikus) landraces as revealed by ISSR marker. Afr. J. Biotechnol. 6, 1460–1468.

Ford, R., Pang, E.C.K., Taylor, P.W.J., 1997. Diversity analysis and species identification in *Lens* using PCR generated markers. Euphytica 96, 247–255.

Furman, B.J., Coyne, C., Redden, B., Sharma, S.K., Vishnyakova, M., 2009. Genetic resources: collection, characterization, conservation and documentation. In: Eriskine, W., Muehlbauer, F., Sarker, A., Sharma, B. (Eds.), The Lentil: Botany, Production and Uses. CABI, Wallingford, pp. 64–75.

Global Crop Diversity Trust (GCDT), 2016. Global Strategy for the Ex-Situ Conservation of Lentil (*Lens* Miller). Available from: http://www.croptrust.org/documents/web/LensStrategy-FINAL-3Dec08.pdf.

Kumar, S., Gupta, S., Chandra, S., Singh, B.B., 2003. How wide is the genetic base of pulse crops. In: Ali, M., Singh, B.B., Kumar, S., Dhar, V. (Eds.), Pulse in New Perspective, Indian Society of Pulse Research. Indian Institute of Pulses Research, Kanpur, India, pp. 34–45.

Kumar, S., Chaudhary, A.K., Rana, K.S., Sarker, A., Singh, M., 2018. Biofortification potential of global wild annual lentil core collection. PLoS ONE 13, e0191122.

Liu, J., Guan, J.P., Xu, D.X., Zhang, X.Y., Gu, J., Zong, X.X., 2008. Genetic diversity and population structure in lentil germplasm detected by SSR markers. Acta Agron. 34, 1901–1909.

Mayer, M.S., Soltis, P.S., 1994. Chloroplast DNA phylogeny of *Lens*: origin and diversity of the cultivated lentil. Theor. Appl. Genet. 87, 773–781.

Redden, B., Maxted, N., Furman, B., 2007. Lens bio-diversity. In: Yadav, S.S., McNeil, D., Stevenson, P.C. (Eds.), Lentil: An Ancient Crop for Modern Times. Springer, Dordrecht, Netherlands, pp. 11–22.

Roberson, L.D., Erskine, W., 1997. Lentil. In: Sears, F.D., Stapleton, P. (Eds.), Biodiversity in Trust. Cambridge University Press, Cambridge, pp. 128–138.

Smartt, J., 1990. Grain Legums: Evolution and Genetic Resources. Cambridge University Press, Cambridge.

Sultana, T., Ghafoor, A., 2008. Genetic diversity in ex-situ conserved *Lens culinaris* for botanical descriptors, biochemical and molecular markers and identification of landraces from indigenous genetic resources of Pakistan. J. Integr. Plant Biol. 50, 484–490.

Zohary, D., 1999. Monophyticvs polyphetic origin of the crops on which agriculture was founded in the north east. Genet. Resour. Crop. Evol. 46, 133–142.

Chapter 2

Origin, Distribution, and Gene Pools

Aqeel Hasan Rizvi*, Muraleedhar Askit[†], Ashutosh Sarker*, Harsh Kumar Dikshit[†], Prachi Yadav[†]

International Center for Agricultural Research in the Dry Areas, South Asia & China Regional Program, New Delhi, India, [†]Indian Agricultural Research Institute, Division of Genetics, New Delhi, India

2.1. INTRODUCTION

The lentil (*Lens culinaris* ssp. *culinaris*) is an ancient and early domesticated legume that continues to play an important role in human and animal diets, and in modern agriculture. The lentil plant was given the scientific name *Lens culinaris* in 1787 by a German botanist and physician, Medikus (Cubero, 1981; Sehirali, 1988; Hanelt, 2001). In different parts of the world, the lentil is known by various names. The most common names are *Lentil* (English), *Adas* (Arabic), *Mercimek* (Turkey), *Messer* (Ethopia), *Masser* or *Massur* (India), *Heramame* (Japanese), *Mangu* or *Margu* (Persian), and *Masura*, *Renuka*, *Mangalaya* (Sanskrit). It is the fourth most important legume crop after beans (*Phaseolus vulgaris* L.), peas (*Pisum sativum*), and chickpeas (*Cicer arietinum*), with an area of 5.5 m ha and production of 6.32 m tons in 2016 (FAO, 2017). Worldwide, production has increased over the last few decades (FAO, 2017). It is an important crop in West Asia, the Indian Subcontinent, Ethiopia, North Africa, Southern Europe, South and North America, and in Oceania. Among the main producers, production has been trending upwards in Canada, the United States (US), Australia, and China, but has been relatively stable in India, Turkey, Syria, Iran, Nepal and Bangladesh. Canada and the United States produce mainly green cotyledon lentils, whereas the rest of the world produces lentils mostly with red cotyledon. Internationally, the trade lies in small-seeded, red cotyledon lentils, which are dominated by Australia, Canada and Turkey, whereas the market in the large-seeded green lentil is held by Canada and the United States. Lentils (*Lens culinaris* Medik) are mainly grown for grains, used as dhal (whole or dehulled), and in various other preparations like lentil soup or deep-fried and eaten as snack. Its composition and nutritional quality make it an important crop, especially in the developing world. It is a rich source of protein (24%–28%) with an abundance of

Lentils. https://doi.org/10.1016/B978-0-12-813522-8.00002-9

lysine, which makes it a good supplement with cereals for balancing the human diet. Its seed provides minerals and vitamins for human nutrition (Sarker et al., 2017) and straw for animal feed. Therefore, it is an important dietary source of energy, protein, carbohydrates, fiber, minerals, vitamins, and antioxidant compounds, as well as diverse nonnutritional components like protease inhibitors, tannins, oligosaccharides, and phytic acid. The lentil plant has the ability to fix atmospheric nitrogen and carbon sequestration, improving soil health.

In India, it is mostly grown in northern and central parts during the winter season as a rain-fed crop after rice, maize, pearl millet, or *rainy season* fallow. It is mainly cultivated in Uttar Pradesh, Madhya Pradesh, Chhattisgarh, Jharkhand, Bihar, and West Bengal. These states together contribute 85% of the area and 90% of the lentil production. In north-eastern parts of the country, it is also cultivated as a *relay* crop with rice, in which seeds of lentils are broadcasted into the standing crop of rice just before its harvest. It is grown on a wide range of soils from light loamy sand to heavy clay soil in the northern parts and moderately deep black soils in Madhya Pradesh and Maharashtra. In south India, the lentil is grown in pockets under conserved moisture after *Kharif* rice in the Belgaum district in Karnataka and has a unique demand in Maharashtra.

The sowing of lentil is popular for mono and sequential cropping, intercropping, mixed cropping, and relay cropping in various countries. In India, Pakistan, Bangladesh, and Nepal, a rice-lentil system is more common, but its cultivation is also done after maize, cotton, sorghum, and pearl millet. In eastern India, the broadcasting of lentil seed in standing rice crops about 15 days before the harvest gives significantly higher grain yield than lentils sown after the harvest of rice. The inclusion of lentils in various cropping systems improves the physical properties of soil and increases the yield of the succeeding cereal crop due to biological nitrogen fixation and other rotational effects.

Lentils are grown as a cool weather or winter crop in the semi-arid tropics, cultivated from sea level to 3800 m. The crop is not suited for the humid tropics. Lentils may survive on a wide range of soils from light loams and alluvial to black cotton soils, and are best on clay soils; they tolerate moderate alkalinity. Salt tolerance is higher during germination than during subsequent development. Lentil seed germinates between 18°C and 21°C, but may germinate at temperature above freezing point. An optimum temperature for growth and yields are around 24°C and temperatures above 27°C are harmful. Waterlogging is more damaging than drought. Lentils are quantitative long-day plants, some cultivars tending to be day-neutral. It is reported to require environments ranging from cool temperate steppe to wet through subtropical dry to moist forest life zones. It tolerates annual precipitation of 2.8–24.3 dm annual mean temperature of 6.3–27.3°C, and pH of 4.5–8.2 (Kay, 1979; Duke, 1981).

2.2. HISTORY

The history of the lentil is as old as agriculture (Helbaek, 1963). It has been cultivated along with wheat, barley, peas, and flax. The carbonized remains of the lentil date back to 11,000 BC from Greece's Franchthi cave, which are the oldest known remains. The ancient Greeks enjoyed lentils as soups and used it to make bread. Pliny has given a detailed account of growing of the crop from seeds, its medicinal properties, and use for various remedies. The famous Apicius also recorded several recipes for lentils. Small-seeded (2–3 mm) types were found at Tell Mureybit in Syria dating to 8500–7500 BC (Zohary, 1972; Hansen and Renfrew, 1978). Lentil remains have been found in Neolithic, aceramic farming villages which were occupied in the 7th millennium BC in the Near-East arc (Helbaek, 1959). The type of agriculture around these lentils cannot be determined, as during this period, small-seeded cultivated lentils could not be differentiated from wild lentil seeds. In an archeological site in northern Israel, the presence of a large storage of lentils clearly established that, by 6800 BC, lentils were a part of farming. Carbonized lentil seeds have been recovered from widely dispersed places such as Tell Ramand in Syria (6250–5950 BC), acreamic Beidha in Jordan, ceramic Hacilar in Turkey (5800–5000 BC), and Tepe Sabz in Iran (5500–5000 BC) (Van Zeist and Bottema, 1971; Helbaek, 1970). In Greece, lentils dating back to 6000–5000 BC have been found in Neolithic settlements such as Argissa-Magula Tessaly (Hopf, 1962) and Nea Mikomedeia, Macedonia (Renfrew, 1969; Van Zeist and Bottema, 1971) and in the same period lentil remains were also seen in Egypt (Matmur, El Omari late 4th millennium, Helbaek, 1963). The archeobotanical remains of lentils have been found in the excavations of the Harappan civilization covering the period of 3300–1300 BC (Sandhu and Singh, 2007). They are mentioned in the Bible's first chapter, Genesis, in the story of Esau, who lost his birthright over a dish of lentils (Genesis 25: 30–34).

2.3. ORIGIN

Understanding of origin of any biological species encounters three main problems: (i) where and when the biological species originated; (ii) where and when that species became a crop; and (iii) how it has evolved as a crop (Cubero et al., 2009). Cubero, in 1981, with the help of archeological data, briefed the places of evidence. The oldest remains of wild lentils were found in Hacilar in Turkey, Ramad in Syria, Jarmo in Iraq, Jericho in Palestine, Beidha in Jordan, and Ali Kosh in Iran, dated around 9000 BP, in aceramic Neolithic layers. Even older remains were found in Mureybit in Syria, c. 10,500 BP. The oldest Greek lentils were dated around 8000 BP, then Central Europe in 5000–7000 BP, from classical Neolithic to early Bronze; in both regions there are some doubtful *nigricans* seeds. Late arrivals were those of India during 3000–4000 BP and Western Europe countries like France, Germany in 3000–3500 BP. Egypt also shows a later arrival (c. 5000 BP) than in Greece and Central Europe, but conditions in the Nile Delta are not favorable for preserving agricultural remains. Archeological data has revealed the existence of both wild and cultivated lentils in the Near East region, which is evidently the place of origin of the crop.

Ladizinsky (1999) suggested that considerable polymorphism was found to exist in the wild accessions of ssp. *orientalis*. However, three accessions from eastern Turkey and northern Syria were shown to share these characteristics with the cultigen, and therefore, can be regarded as members of the genetic stock from which the cultivated lentil was domesticated. Based on founder effects revealed by chromosome and DNA polymorphisms, as well as evidence from domestication traits and species diversity that the lentil was domesticated once or only a few times (Zohary, 1999). Lev-Yadun et al. (2000) argued that the lentil might have originated in a place close to or overlapping the area in the Fertile Crescent where einkorn and emmer wheats were domesticated.

In contrast, Barulina (1930), based on the Vavilovian criterion to define centers of origin, suggested the eastern border of southwest Asia as a possible center of origin of the cultivated lentil, as the region between Afghanistan, India, and Turkistan (i.e., the Himalaya-Hindu Kush junction) showed the highest proportion of endemic varieties. She also noticed that the area of distribution of wild lentils did not overlap much with that of the domesticated ones, but she maintained her idea considering that the eastern part of the *orientalis* area of distribution reaches Turkmenia. Thus, the overlap between the wild ancestor and the cultigen and the archeological remains connecting both are *sine qua non* conditions to establish such a center of origin. Three groups are restricted to very concrete areas: *pilosae* to the Indian subcontinent, *aethiopicae* to Ethiopia and Yemen, and *subspontanea* to the Afghan regions closest to the Indian subcontinent. All three have very small and dark-colored seeds, violet flowers, few flowers per peduncle, calyx teeth much shorter than the corolla, few leaflets per leaf, and dwarf plants. Each one of these three shows a distinct character: *pilosae*—a strong pubescence; *subspontanea*—very dehiscent pods, being purple colored before maturity; and *aethiopicae*—pods with a characteristic elongated apex.

These particular traits are shown together with a cluster of primitive traits closely related to *orientalis*. The other three *microsperma* groups (*europeae*, *asiaticae*, and *intermediae*) and *macrosperma* are rather cosmopolitan and are clearly intermixed, even when *intermediae* has a rather restricted area. Seeds are variable in size, but in general, they are wider than 4 mm. They have more flowers per peduncle, more leaflets per leaf, and the calyx teeth are equal to or larger than the corolla. White flowers are common, with seeds being diverse in color. The subspontanea overlaps only with *orientalis*, and aethiopicae with *ervoides*; pilosae does not overlap with any wild lentil. All other microsperma as well as macrosperma lentils overlap to a greater or lesser extent with all the known wild lentils. On the contrary, no lentils have been found in the sites dating back to the seventh millennium BP in Turkmenia. The high degree of endemism that exists in the Afghanistan-Indian-Turkmenian area is better explained, as in all other species, by an intense genetic drift, typical of highly diversified environments, coupled with artificial selection carried out by very diverse human populations, with drastic genetic fixation and losses providing secondary centers of diversity.

In conclusion, the area from western Turkey to northern Iraq contains not only all the wild lentils but also "lentoid" characters such as flattened pods and seeds present in other species like *V. montbretii* (possible bridge between *Vicia* and *Lens*) and *V. lunata*. The known evidence suggests that from southern Turkey to northern Syria region is the most likely place of lentil domestication. Some populations of orientalis were unintentionally subjected to automatic selection here, leading to a new crop, *Lens culinaris*.

2.4. TAXONOMY, MORPHOLOGY AND FLORAL BIOLOGY

The genus *Lens* Miller is a member of the tribe *Vicieae*, subfamily *Papilionaceae*, family *Leguminosae*. Beside *Lens*, three other genera are included in the *Vicieae*: *Vicia* L., *Lathyrus* L., and *Pisum* L. From a morphological point of view, a range exists between the genera *Lens* and *Vicia*. However, *Lens* is a much smaller genus, characterized by an annual growth habit, small flowers, calyx deeply divided into subulate, subequal teeth, and a broadly rhomboid compressed legume with one or two orbicular flattened seeds. The genus *Lens* comprises seven taxa in six species (Ferguson, 1998; Ferguson et al., 2000). *Lens orientalis* is the presumed progenitor of *Lens culinaris*, and the two species are crossable and produce fully fertile progeny (Muehlbauer et al., 2006). According to crossability, phenetic relations, and chromosomal diversity, Ladizinsky and Abbo (1993) suggested two biological species in the genus Lens: *Lens culinaris* and *Lens nigricans*, with a few subspecies. However, additional information now indicates that some of the proposed subspecies are species in their own right. In 1997, two new species were recognized in genus Lens. *Lens tomentosus* was separated from *Lens culinaris* subsp. *orientalis* on the basis of its tomentose, as opposed to puberulent, pods, and a relatively small asymmetrical chromosome which bears a minute satellite (Ladizinsky, 1997). *Lens lamottei*, originally described by Czefranove (1971), was found to be the same as a differentiated cytotype identified within *Lens nigricans* by Ladizinsky et al. (1983, 1984) and is now recognized as a separate taxon (Van Oss et al., 1997). Thus, as a result of combined evidence of crossability, phenetic relations, and morphological markers (Ferguson and Erskine, 2001; Ferguson et al., 2000), the genus *Lens* consists of the six species. From the standpoint of crossability for use in breeding, the *Lens* species can be divided into three groups: *L. culinaris* and *L. odemensis* make up the primary genepool, *L. ervoides* and *L. nigricans* belong to the secondary genepool, and *L. lamottei* and *L. tomentosus* belong to the tertiary genepool (Muehlbauer and McPhee, 2005). Crosses between members of the different genepools generally fail, because the hybrid embryos abort. However, embryo rescue has been used successfully to obtain viable hybrids between groups (Ladizinsky et al., 1985). The basic chromosome number of the genus *Lens* is $n = 7$. All the *Lens* species share more or less the same karyotype, which includes three pairs of metacentric, or submetacentric chromosomes, a pair of metacentric chromosome with a secondary constriction very close to the centromere, and three pairs of acrocentric chromosomes (Ladizinsky and Abbo, 1993).

Varietal identification was realized by choosing convenient, nongeographical, and sometimes utilitarian characteristics. Lentil in taxonomy is as follows (Anonymous, 2012):

Kingdom: *Plantae*-Plants,
Subkingdom: *Tracheobionta*-Vascular plants,
Superdivision: *Spermatophyta*-Seed plants,
Division: *Magnoliophyta*-Flowering plants,
Class: *Magnoliopsida*-Dicotyledons,
Subclass: *Rosidae*,
Order: *Fabales*,
Family: *Fabaceae*-Pea family,
Genus: *Lens* Mill.-lentil,
Species : *Lens culinaris* Medik.-lentil.

Taxonomic analyses based on morphological and/or biochemical markers ranged from four species in 1979, namely, *L. culinaris*, *L. orientalis*, *L. nigricans*, and *L. ervoides*; to two species in 1984: *L. culinaris* (with subspecies *culinaris* (the cultigen), *orientalis* and *odemensis*) and *L. nigricans* (with ssp. *nigricans* and *ervoides*); again to four species in 1993, that is, *L. culinaris* (ssp. *culinaris* and *orientalis*), *L. odemensis*, *L. nigricans*, and *L. ervoides*; to also four species in the year 2000, that is, *L. culinaris* (ssp. *culinaris*, *orientalis*, *tomentosus*, and *odemensis*), *L. nigricans*, *L. ervoides*, and *L. lamottei*; and six species after 2000, which are *L. culinaris* (ssp. *culinaris* and *orientalis*), *L. odemensis*, *L. tomentosus*, *L. nigricans*, *L. ervoides*, and *L. lamottei*. The contradictory results of different studies are a consequence of the evolutionary process itself. *Lens* species share many common structural and biochemical features, and the distances obtained in different studies are a function of the origin of the accessions involved in them, as well as of the particular characters and molecular markers chosen. Subtle differences can place accessions in distinct clusters. It is not a mistake of the experimental method; rather, it is a consequence of a radiating evolution (Cubero et al., 2009).

The lentil is a diploid species ($2n = 14$) (Muehlbauer, 1991). It is a self-pollinating annual species with a haploid genome size of an estimated 4063 Mbp (Arumuganathan and Earle, 1991). The botanical features of *Lens culinaris* (cultivated lentil) can be described as annual, bushy herb, slender almost erect or suberect, much-branched, softly hairy; stems slender, angular, 15–75 cm height (Duke, 1981; Muehlbauer et al., 1985). Ten to sixteen leaflets are subtended on the rachis (40–50 mm); upper leaves have simple tendrils, while lower leaves are mucronate (Muehlbauer et al., 1985). "The leaves are alternate, compound, pinnate, usually ending in a tendril or bristly; leaflets 4–7 pairs, alternate or opposite; oval, sessile, 1–2 cm long; stipules small, entire; stipels absent; pods oblong, flattened, or compressed, smooth, to 1.3 cm long, 1-2-seeded; seed biconvex, rounded, small, 4–8 mm × 2.2–3 mm, lens-shaped, green, greenish-brown, or light red speckled with black; the weight of 100 seeds range from 2 to 8 g; cotyledons red, orange, yellow, or green, bleaching to yellow, often showing through the testa, influencing its apparent color" (Kay, 1979; Duke, 1981; Muehlbauer et al., 1995). Flowers are small, pale blue, purple, white, or pink, in axillary 1–4-flowered racemes; 1–4 flowers are borne on a single peduncle, and a single plant can produce from 10 to 150 peduncles each being 2.5–5 cm long (Muehlbauer et al., 1985). Flowering proceeds acropetally. The size of seeds increases from the types grown in eastern regions to western types. Two types, namely, macrosperma, found mainly in the Mediterranean region and the New World (seed size ranging from 6 to 9 mm in diameter and yellow cotyledons with little or no pigmentation), and microsperma (2–6 mm with red-orange or yellow cotyledons) found on the Indian subcontinent and in the Near East and East Africa, respectively, are known (Hawtin et al., 1980; Muehlbauer et al., 1985). The first one includes the Chilean or yellow cotyledon types, while the latter includes the small-seeded Persian or red cotyledon lentils (Kay, 1979). Germination is hypogeal, and this keeps the developing seedlings below ground level, which reduces the effects of freezing and other desiccating environmental conditions (Muehlbauer et al., 1985).

2.5. DISTRIBUTION

Being first cultivated in the Fertile Crescent, then moving to Greece and Central and Western Europe along the Danube, to the Nile Delta, and eastward to India, lentils followed the spreading route of the Neolithic agricultural techniques (Harlan, 1992). From Greece, lentils found their way to Central Europe through the Danube. Since the early Neolithic, there are lentil archeological remains in the Iberian Peninsula, the oldest ones corresponding to the eastern part of the Peninsula. In particular, in the cave, "de les Cendres," there are remains of several crops (*Triticum monococcum*, *Triticum dicoccum*, *Triticum aestivum*, barley, peas, grass peas, lentils, and faba beans, that is, a typical Near-East crop complex) dated by ^{14}C to 7540 ± 140 BP (Cubero et al., 2009). The archeological lentil findings in the Iberian Peninsula, conferring to the seed size, fall within the range of the *microsperma* sizes (Buxo, 1997). *Intermediae* forms reached Sicily, and *asiaticae* forms reached Sardinia, Morocco, and Spain, suggesting the arrival in these countries of lentil stocks from Central Europe, or from the route of the isles (Cubero et al., 2009). De Candolle (1882) suggested that lentils may have reached the cradle of the Indo-European people after the Greek ancestors split on linguistic grounds: Greek for lentil is *phakos*, but *lens* in Latin, *lechja* in Illyrian, and *lenzsic* in Lithuanian. It is likely that the ancient Greeks took the word from the aboriginal Mediterranean populations they conquered. Central Russia, then Siberia, was more likely reached from the western coast of the Black Sea or from the Danube valley rather than from Mesopotamia or Central Asia. The arrival of lentils to the Nile Delta had to be much earlier than the archeological remains suggest due to its proximity, both geographical and cultural, to the Fertile Crescent; but the Delta environment is not helpful in preserving organic materials. Ethiopia was probably reached from the Arabian coast (at that time it was the *Arabia Felix*, more humid and fertile than now), rather than by the Nile. Lentils likely travelled in the Near East complex (along with barley, wheat, chickpeas, faba beans, etc.), established in the Ethiopian highlands since primitive times, and have been evolving since then in isolation, producing much endemism that allowed Vavilov to place the centers of origin of crop plants that only arrived there by farming. Indeed, *aethiopicae* exhibits very primitive characters, meaning that lentils probably also arrived at a very primitive stage of domestication (Cubero et al., 2009). Lentils did not reach India before 4000 BP, probably carried by the Indoeuropean invasion (De Candolle, 1882) through Afghanistan, but again there is the problem of availability of archeological findings. Primeval introduction was probably performed by very small samples of a common origin as the variability found in the Indian subcontinent in the local landraces is very limited, in spite of being the largest lentil-growing region in the world; the asynchrony in flowering of the local *pilosae* ecotype, probably a consequence of a long reproductive isolation period, is now being broken by plant breeding methods in order to widen the genetic base available (Erskine et al., 1998).

2.6. GENE POOL

The evidence produced on the basis of crossability between the crop species and the subsequent cytogenetic, phylogenetic, and molecular studies has been used to characterize the crop species into different gene pools (Harlan and De Wet, 1971). The wild relatives can be divided into primary, secondary, and tertiary gene pools, according to their relatedness to *L. culinaris* ssp. *culinaris* and their ability to produce fertile hybrids when intercrossed with cultivated lentils. The species of lentils have been grouped into primary (*Lens culinaris* ssp. *culinaris*, *L. culinaris* ssp. *orientalis*, *L. odemensis*), secondary (*L. ervoides*, *L. nigricans*), and tertiary gene pools (*L. lamottei*, *L. tomentosus*) (Ladizinsky et al., 1984; Ladizinsky, 1999; Muehlbauer and McPhee, 2005). Hancock (2004) grouped lentil species into three groups only on the basis of crossing, and placed intercrossable species *L. nigricans*, *L. ervoides*, and *L. lamottei* into a secondary pool or Group II. Crossing of Group I species with *L. nigricans* produces nonviable seeds in hybrids due to irregular meiosis (Ladizinsky et al., 1984, 1985). However, the use of embryo rescue can produce viable seed in hybrids derived from crossing between *L. culinaris* and *L. ervoides* (Ladizinsky et al., 1985). They also put *L. tomentosus* in Group III or the tertiary gene pool as a single species group, and this species does not produce viable seed in hybrids derived by crossing with other group of species. Schaefer et al., 2012 noted that the genus *Lens* ($2n = 14$) is phylogenetically comes under the tribe "Vicieae," which are cool-season legumes belonging to subfamily Papilionoideae of family Fabaceae. The genus *Lens* is comprised of seven closely related taxa, namely,

L. culinaris, *L. orientalis*, *L. tomentosus*, *L. odemensis*, *L. lamottei*, *L. ervoides*, and *L. nigricans*. Most of the past taxonomic studies based on crop morphology, cytogenetics, hybridization studies, and/or molecular markers have frequently disagreed with respect to classification at the species and subspecies level. The most recent classification by Kole et al. (2011), identified seven taxa grouped into four species, namely, *L. culinaris* ssp. *culinaris*, *L. culinaris* ssp. *orientalis*, *L. culinaris* ssp. *tomentosus*, *L. culinaris* ssp. *odemensis*, *L. ervoides*, *L. lamottei*, and *L. nigricans*. Despite the taxonomic reorganizations, all studies generally agreed that *L. culinaris* ssp. *orientalis* is the most closely related wild progenitor of *L. culinaris* ssp. *culinaris*, while *L. nigricans* is the most distant relative.

For genetic enhancement by generating new variability, valuable genes recognized in the primary gene pool have readily been used for crop improvement. Nevertheless, for incidence, the useful genes have been observed much more in the species of the secondary and tertiary gene pools (Collard et al., 2001; Tullu et al., 2006). Accordingly, more efforts are required on deployment of novel techniques for using these gene pools for the enhancement of lentils. Fratini and Ruiz (2006) suggested that a hybrid between *L. culinaris* ssp. *culinaris* and *L. nigricans*, *L. ervoides*, and *L. odemensis* developed through embryo rescue can be viable with a rate of 3%–9%. Based on these observations, *L. odemensis* has been considered to be a member of secondary gene pool, and *L. nigricans* and *L. ervoides* were classified in the tertiary gene pool (Ladizinsky, 1993). Further, Cubero et al. (2009) also suggested placing *L. odemensis* in the secondary gene pool, while *L. nigricans* and *L. ervoides* can also be part of secondary gene pool as hybrids which could be obtained by means of embryo rescue. They also suggested that there is a need to study hybridization in order to establish place of *L. tomentosus* and *L. lamottei* in secondary or tertiary gene pool. Some early studies, based on cross compatibility and cytogenetic evidence placed *L. orientalis* in the primary gene pool, while the secondary gene pool consisted of *L. odemensis*, *L. ervoides*, and *L. nigricans*. Ladizinsky and Muehlbauer 1993 noted that the gene pool placement of two more recently identified species *L. lamottei* and *L. tomentosus* is inconsistent among studies. These two species were placed in the secondary gene pool by Muehlbauer and McPhee (2005). Fratini and Ruiz (2006) suggested that *L. ervoides* and *L. nigricans* should be placed in the tertiary gene pool. Nonetheless, the gene pools proposed by all these studies are not in concordance with the species classification from the study of Ferguson et al. (2000), suggesting more work is needed to clarify these relationships.

Most recently Wong et al. (2015) used next-generation, sequencing-based genotyping methods for screening of the lentil germplasm collection using a two-enzyme GBS approach to clarify the doubts on gene pool classification in lentil. They constructed two 96-plex GBS libraries with a total of 60 accessions where some accessions had several samples, and each sample was sequenced in two technical replicates. They detected a total of 266,356 genome-wide SNPs by automated GBS pipeline. Based on haplotype information, they filtered low quality and redundant SNPs, and constructed a maximum-likelihood tree using 5389 SNPs. The phylogenetic tree grouped the germplasm collection into their respective taxa with strong support. Based on phylogenetic tree and STRUCTURE analysis, they identified four gene pools, namely, *L. culinaris/L. orientalis/L. tomentosus*, *L. lamottei/L. odemensis*, *L. ervoides*, and *L. nigricans*, which form primary, secondary, tertiary, and quaternary gene pools, respectively.

2.7. DOMESTICATION

The lentil was utilized by man during the early stages of the Neolithic Revolution. Remains of lentil seeds in archeological digs suggest that it was one of the first plants to be exploited by man (Zohary and Hopf, 1988). The oldest seed remains come from the Middle East, hence, the prevailing idea that lentil domestication occurred here, together with that of other pulses and cereals. Attempts to identify the place of domestication based on the occurrence of carbonized lentil seeds in archeological sites are also not without complications, because there is no way to distinguish between wild and cultivated forms found in these remains. By accepting *L. nigricans* as the wild progenitor of lentils, Renfrew (1969, 1973) concludes southern Europe was the place where the lentil evolved. But Zohary (1972) and Williams et al. (1974), favoring Barulina's (1930) idea that lentil originated from *L. orientalis*, interpreted the occurrence of carbonized lentil seeds in Neolithic settlements in the Middle East as a sign that the lentil was domesticated throughout the Fertile Crescent. Archeological studies, have confirmed the presence of lentil in the Turkey-Syria-Iraq region as far back as 8500–600 BC. This region probably played an important role in lentil domestication and the further spread to the Nile, Greece, Central Europe, and eastward to South Asia (Nene, 2006). At the same time, lentils also spread to Ethopia, Afghanistan, India, Pakistan, China, and later to the New World, including Latin America (Cubero, 1981, Duke, 1981, Ladizinsky, 1979a). Bahl et al. (1993) suggested that lentils' domestication was probably the oldest among grain legumes. Domestication may have started with the selection of plants from wild species that retain their seeds in pods before harvesting, and continuous selection for large seed size. Two main traits for lentils were involved in domestication, pod dehiscence and seed dormancy, and both are reported to be under the control of a single recessive gene. A third major trait, seed size, appears to be under more complex control (Sonnante et al., 2009). Many authors consider seed dispersal to have been the first character entity selected by man, because of its consequences on harvest effectiveness, thus yield, and seed increase efficiency (Zohary, 1999). Certainly, seed size increase followed later, because archeological relics show no seed size increase until the Bronze Age (Fuller, 2007). To establish how and when domestication took place, evidence tends to suggest that cultivation was carried out by man far before domestication traits were fixed (Pringle, 1998; Balter, 2007). The most possible pattern is that man consciously targeted those cultivated wild species which were carrying the genes targeted by domestication. Therefore, consider domestication as a guided evolution, in which natural selection was substituted with agricultural pressure. The evidence that domestication may not be a sudden process might also come from the observation that seed size increase is surely successive to the establishment of a real crop (Sonnante et al., 2009).

REFERENCES

Anonymous, 2012. Taxonomical Database. United States Department of Agriculture, Natural Resources Conservation Service. Retrieved from: http://plants.usda.gov.

Arumuganathan, K., Earle, E.D., 1991. Nuclear DNA content of some important plant species. Plant Mol. Biol. Report. 9, 208–218.

Bahl, P.N., Lal, S., Sharma, B.M., 1993. An overview of the production and problems in southeast Asia. In: Erksine, W., Saxena, M.C. (Eds.), Lentil in South Asia-Proceedings of the Seminar on Lentils in South Asia, ICARDA, Aleppo, Syria. pp. 1–10.

Balter, M., 2007. Seeking agriculture's ancient roots. Science 316, 1830–1835.

Barulina, H., 1930. Lentils of the USSR and other countries. Bull. Appl. Bot. Genet. Plant Breed. 40, 265–304.

Buxo, R., 1997. Arqueologia de las Plantas. Critica, Barcelona, Spain.

Collard, B.C.Y., Ades, P.K., Pang, E.C.K., Brouwer, J.B., Taylor, P.W.J., 2001. Prospecting for sources of resistance to ascochyta blight in wild *Cicer* species. Aust. J. Plant Pathol. 30, 271–276.

Cubero, J.I., 1981. Origin, domestication and evolution. In: Webb, C., Hawtin, G.C. (Eds.), Lentils. Commonwealth Agricultural Bureau, Slough, UK, pp. 15–38.

Cubero, J.I., Perez de la Vega, M., Fratini, R., Sharma, B., 2009. Origin, phylogeny, domestication and spread. In: Erskine, W., Muehlbauer, F.J., Sarker, A. (Eds.), The Lentil: Botany, Production and Uses. CABI, Oxfordshire, pp. 13–33.

Czefranove, Z., 1971. Novosti Systematischeski Vyssich Rastenii. vol. 8. 184–191.

De Candolle, A.P., 1882. Origins of Cultivated Species. Hafner, London, UK. (Reprinted 1967).

Duke, J.A., 1981. Handbook of Legumes of World Economic Importance. Plenum Press, New York52–57.

Erskine, W., Chandra, S., Chaudary, M., Malik, I.A., Sarker, A., Sharma, B., Tufail, M., Tyagi, M.C., 1998. A bottleneck in lentil: widening the genetic base in South Asia. Euphytica 101, 207–211.

FAO, 2017. Faostat, Fao Statistical Database. Retrieved from: http://www.fao.org/faostat/en/#data/QC.

Ferguson, M., 1998. Studies of Genetic Variation Within the Genus Lens. (PhD thesis) School of Biological Sciences, University of Birmingham, Birmingham, UK.

Ferguson, M.E., Erskine, W., 2001. Lentils (*Lens* L.). In: Maxted, N., Bennett, S.J. (Eds.), Plant Genetic Resources of Legumes in the Mediterranean. Kluwer Academic Publishers, Dordrecht, Netherlands, pp. 132–157.

Ferguson, M.E., Maxted, N., Van Slageren, M., Robertson, L.D., 2000. A reassessment of the taxonomy of *Lens* Miller (Leguminosae, Papilionoideae, Vicieae). Bot. J Linnean Soc. 133, 41–59.

Fratini, R., Ruiz, M.L., 2006. Interspecific hybridization in the genus *Lens* applying *in vitro* embryo rescue. Euphytica 150, 271–280.

Fuller, D.Q., 2007. Contrasting patterns in crop domestication rates: recent archaeobotanical insights from the old world. Ann. Bot. 100, 903–924.

Hancock, J.F., 2004. Plant Evolution and the Origin of Crop Species, second ed. CABI, Oxfordshire, UK.

Hanelt, P., 2001. *Lens* Mill. In: Hanelt, P. (Ed.), Mansfeld's Encyclopedia of Agricultural and Horticultural Crops. vol. 2. pp. 849–852.

Hansen, J., Renfrew, J.W., 1978. Paleolithic-Neolithic seed remains at Franchthi cave, Greece. Nature 71, 349–352.

Harlan, J., 1992. Crops and Man. American Society of Agronomy, Madison, Wisconsin, USA.

Harlan, J.R., De Wet, J.M.J., 1971. Towards a rational classification of cultivated plants. Taxon 20, 509–517.

Hawtin, G.C., Singh, K.B., Saxena, M.C., 1980. Some recent developments in the understanding and improvement of *Cicer* and *Lens*. In: Summerfield, R.J., Bunting, A.H. (Eds.), Advances In Legume Science. pp. 613–623.

Helbaek, H., 1959. Domestication of food plants in old world. Science 130, 365–372.

Helbaek, H., 1963. Late cypriote vegetable diet in Apliki. Act Nstit Athen Reg. Sueciae Ser. 4, 171–186.

Helbaek, H., 1970. The plant industry in Hacilar. In: Mellart, J. (Ed.), Excavations at Hacilar. vol. 1. Edinburgh University Press, pp. 189–244. Occasional Publications No. 9 of the British Institute of Archaelogy, Ankara,.

Hopf, M., 1962. Bericht Uber die Untersuchung von Samen und Holzkohlenresten von der Argissa-Magula aus den prakermischen bis mittelbronzezeitlichen Schichten. In: Milojcic, V., Boessneck, J., Hopf, M. (Eds.), Die Deutschen Ausgrabungen auf der Argissa-Magula in Thessalien. I. Rudolf Habelt Verlag, Bonn, pp. 101–119.

Kay, D., 1979. Food legumes. Tropical Development and Research Institute (TPI), TPI Crop.

Kole, C., Gupta, D., Ford, R., Taylor, P.J., 2011. Lens-Wild Crop Relatives: Genomic and Breeding Resources. Springer, Berlin Heidelberg127–139.

Ladizinsky, G., 1979. The origin of lentil and its wild gene pool. Euphytica 28, 179–187.

Ladizinsky, G., 1993. Wild lentils. Crit. Rev. Plant Sci. 12, 169–184.

Ladizinsky, G., 1997. A new species of *Lens* from south-east Turkey. Bot. J. Linn. Soc. 123, 257–260.

Ladizinsky, G., 1999. Identification of the lentil wild genetic stock. Genet. Resour. Crop. Evol. 46, 115–118.

Ladizinsky, G., Abbo, S., 1993. Cryptic speciation in *Lens culinaris*. Genet. Res. Crop. Evol. 40, 1–5.

Ladizinsky, G., Braun, D., Muehlbauer, F.J., 1983. Evidence for domestication of *Lens nigricans* (M. Bieb) Godron in S Europe. Bot. J. Linn. Soc. 87, 169–176.

Ladizinsky, G., Braun, D., Goshen, D., Muehlbauer, F.J., 1984. The biological species of the genus *Lens*. Bot. Gaz. 145, 253–261.

Ladizinsky, G., Cohen, D., Muehlbauer, F.J., 1985. Hybridization in the genus *Lens* by means of embryo culture. Theor. Appl. Genet. 70, 97–101.

Lev-Yadun, S., Gopher, A., Abbo, S., 2000. The cradle of agriculture. Science 288, 1602–1603.

Muehlbauer, F.J., 1991. Use of introduced germplasm in cool season food legume cultivar development. In: HL, S., Wiesner, L.E. (Eds.), Use of Plant Introductions in Cultivar Development (Part 2). vol. 20. Crop Sci. Soc. Amer. Social Publication, pp. 49–73.

Muehlbauer, F.J., McPhee, K.E., 2005. Lentil-*Lens culinaris* Medik. In: Singh, R.J., Jauhar, P.P. (Eds.), Genetic Resources, Chromosome Engineering and Crop Improvement, Grain Legumes. Taylor & Francis, Boca Raton, FL, pp. 219–230.

Muehlbauer, F.J., Cubero, J.I., Summerfield, R.J., 1985. Lentil (*Lens culinaris Medik.*). In: Summerfield, R.J., Roberts, E.I.I. (Eds.), Grain Legume Crops. vol. 8. Collins, Grafton Street, London, UK, pp. 266–311.

Muehlbauer, F.J., Kaiser, W.J., Clement, S.L., Summerfield, R.J., 1995. Production and breeding of lentil. Adv. Agron. 54, 283–332.

Muehlbauer, F.J., Cho, S., Sarker, A., McPhee, K.E., Coyne, C.J., Rajesh, P.N., Ford, R., 2006. Application of biotechnology in breeding lentil for resistance to biotic and abiotic stress. Euphytica 147, 149–165.

Nene, Y.L., 2006. Indian pulses through the millennia. Asian Agric. Hist. 10, 179–202.

Pringle, H., 1998. Neolithic agriculture: the slow birth of agriculture. Science 282, 1446.

Renfrew, J.M., 1969. The archaeological evidence for the domestication of plants: methods and problem. In: Ucko, P.J., Dimbleby, G.W. (Eds.), The Domestication and Exploitation of Plants and Animals. Aidine, Chicago.

Renfrew, J.M., 1973. Palaeoethnobotany. Columbia Univ. Press, New York.

Sandhu, J.S., Singh, S., 2007. In: Yadav, S.S., McNeil, D., Stevenson, P.C. (Eds.), History and Origin in Lentil: An Ancient Crop for Modern Times. Springer, pp. 1–9.

Sarker, A., Rizvi, A.H., Singh, M., 2017. Genetic variability for nutritional quality in lentil (*Lens culinaris* Medikus subsp. *culinaris*). Legume Res. 1–6.

Schaefer, H., Hechenleitner, P., Santos-Guerra, A., Menezes de Sequeira, M., Pennington, R.T., Kenicer, G., et al., 2012. Systematics, biogeography, and character evolution of the legume tribe Fabeae with special focus on the middle-Atlantic island lineages. BMC Evol. Biol. 12, 250.

Sehirali, S., 1988. Grain legume crops. Ankara University, Faculty of Agricultural Engineering, Ankara, Turkey435.

Sonnante, G., Hammer, K., Pignone, D., 2009. From the cradle of agriculture a handful of lentils: history of domestication. Rendiconti Lincei 20, 21–37.

Tullu, A., Buchwaldt, L., Lulsdorf, M., Banniza, S., Barlow, B., Slinkard, A.E., Sarker, A., Tar'an, T.D., Warkentin, T.D., Vandenberg, A., 2006. Sources of resistance to anthracnose *Colletotrichum truncatum* in wild *Lens* species. Genet. Resour. Crop. Evol. 53, 111–119.

Van Oss, H., Aron, Y., Ladizinsky, G., 1997. Chloroplast DNA variation and evolution in the genus *Lens*. Theor. Appl. Genet. 106, 428–434.

Van Zeist, W., Bottema, S., 1971. Plant husbandry in early neolithic Nea Nikomedeia, Greece. Acta Bot. Neerl. 20, 521–538.

Williams, J.T., Sanchez, A.M.C., Jackson, M.T., 1974. Studies on lentils and their variation in the taxonomy of the species. SABRAO J. 6, 133–145.

Wong, M.M.L., Gujaria-Verma, N., Ramsay, L., Yuan, H.Y., Caron, C., Diapari, M., et al., 2015. Classification and characterization of species within the genus *Lens* using genotyping-by-sequencing (GBS). PLoS ONE 10, e0122025.

Zohary, D., 1972. The wild progenitor and the place of origin of the cultivated lentil *Lens culinaris*. Econ. Bot. 26, 326–332.

Zohary, D., 1999. Monophyletic vs. polyphyletic origin of crops on which agriculture founded in the near east. Genet. Resour. Crop. Evol. 46, 133–142.

Zohary, D., Hopf, M., 1988. Domestication of Plants in the Old World. Clarendon Press, London.

Chapter 3

Genetic Resources: Collection, Conservation, Characterization and Maintenance

Nikhil Malhotra, Sweety Panatu, Badal Singh, Narender Negi, Dayal Singh, Mohar Singh, Rahul Chandora

ICAR-National Bureau of Plant Genetic Resources, Regional Station, Shimla, India

3.1. INTRODUCTION

"Lentil" (*Lens culinaris* ssp. *culinaris* Medikus) is a Latin word, and the German botanist Medikus gave the lentil its scientific name, "*Lens culinaris*," in 1787 (Cubero, 1981; Sehirali, 1988; Hanelt, 2001). The lentil is a true diploid ($2n = 2\times = 14$) (Muehlbauer, 1991), annual, self-pollinating crop with a genome size of 4063 Mbp (Arumuganathan and Earle, 1991). It is the oldest cultivated legume plant (Bahl et al., 1993; Rehman and Altaf, 1994), as old as those of einkorn, emmer, peas, and barley (Harlan, 1992), and domesticated by mankind on the planet since the Neolithic period (Ljustina and Miki'c, 2010). It is the third most important temperate grain legume crop in the world after chickpeas (*Cicer arietinum* L.) and peas (*Pisum sativum* L.) (FAO, 2015). The major lentil-growing countries in the world are India, Turkey, Canada, Australia, the United States, Nepal, China, and Ethiopia. India ranks first in lentil production (1.31 million tons) as well as in lentil-sown area (1.39 million hectares) followed by Turkey, whereas globally, lentil is grown over 4.34 million hectares with annual production and productivity of 4.95 million tons and 1260 kg ha^{-1}, respectively (FAO, 2015). Lentils are produced mostly in marginal areas under rain-fed conditions with a global average yield of 850 kg ha^{-1} (Erskine et al., 2011). It is an annual, bushy, leguminaceous plant providing an ample source of protein, minerals (K, P, Fe, and Zn), and vitamins (Bhatty, 1988). The crop has high lysine and tryptophan content, and its straw is also a valued animal feed (Eriskine et al., 1990). Lentil was domesticated in the Near East in an area called "the cradle of agriculture" about 11,000 BC (Sonnante et al., 2009) from the wild progenitor, *L. culinaris* ssp. *orientalis* (Boiss.) Ponert. The cultivated lentil is reported to have two centers of origin, India and the Middle East (Ladizinsky, 1993), whereas

Lentils. https://doi.org/10.1016/B978-0-12-813522-8.00003-0

wild lentil species may have multiple centers of origin (Pratap et al., 2014; Singh et al., 2014a,b). Further, most of the current lentil varieties possess narrow genetic bases due to bottleneck effects which act as a barrier to achieving desired productivity to satisfy increased food security demand. However, lentil production is also limited by many other factors, including abiotic stresses such as terminal drought, heat stress, low soil fertility, and major biotic stresses such as ascochyta blight (*Ascochyta lentis*), anthracnose (*Colletotrichum truncatum*), stemphylium blight (*Stemphylium botryosum*), rust (*Uromyces viciae-fabae*), fusarium wilt (*Fusarium oxysporum* f. sp. *lentis*), botrytis gray mold (*Botrytis cinerea* and *B. fabae*), and white mold (*Sclerotinia sclerotiorum*) (Sharpe et al., 2013; Kumar et al., 2015). In the last 50 years, the cultivated area of the lentil has been reduced due, in part, to low productions because it has received less attention in breeding programs than other legume crops. Therefore, building and improving an effective lentil breeding program requires utilizing lentil genetic resources (including wild species) in order to develop valuable genotypic and phenotypic variation into the background of elite germplasm (Singh et al., 2014a,b). Hence, germplasm exploration, collection, characterization, conservation, and maintenance are needed for tailoring new plant types adapted to varied agro-ecological conditions to meet the food security needs of the world.

3.2. SPECIES DISTRIBUTION AND GENE POOLS

The wild *Lens* taxa are widely distributed in the Mediterranean region, and it was thought that only the Aegean and Southwest of Turkey overlapped for the distribution of *Lens* taxa (Ferguson et al., 1996). The distribution of the subspecies, *L. culinaris* ssp. *orientalis*, has an Eastern spread from Turkey, and *L. culinaris* ssp. *odemensis* has restricted distribution in the East, spreading from Turkey southwards to Syria and Palestine. *L. culinaris* ssp. *tomentosus* is found in Southeastern Turkey (Ladizinsky, 1997; Ferguson et al., 1998b). However, the species *L. ervoides* has a wide distribution from Spain to Ukraine and south to Jordan. *L. nigricans* grows in diffused small patches on rocky hillsides (Zohary, 1999) and has a Western distribution from Spain to Turkey and south to Morocco (Ferguson et al., 1996), while *L. lamottei* grows well in Morocco (Van Oss et al., 1997). Unfortunately, Turkey, like other Mediterranean countries, is suffering from the rapid loss of many of its invaluable genetic resources. These genetic resources, which have the potential to provide useful diversity for crop breeding efforts, are being eroded primarily due to habitat destruction (Solh and Erskine, 1981).

3.3. GERMPLASM CONSERVATION

Genetic resources are an essential component to meet the future food security needs of the world. They act as a reservoir of useful genes and alleles, which contribute enormously toward the genetic enhancement of crop plants and provide raw material for crop improvement programs. Conserved plant genetic resources are essential to meet the current and future needs of crop improvement programs. The concept of germplasm conservation demands that collection methods initially capture maximum variation, and subsequently, conservation and regeneration techniques minimize losses through time. The germplasm must be collected from diversity-rich areas of the world to enrich these genetic resources. In the last three decades, several studies have been conducted on the collection of lentil species and their variability among the germplasm in the world. These collections have been further utilized in lentil improvement programs (Erskine and Witcombe, 1984; Fikiru et al., 2007). Globally, about 7.4 million germplasm accessions of different crop species have been collected and maintained in over 1750 ex-situ gene banks (FAO, 2010), while legume germplasm stands as the second largest collection (15%) after cereals (45%). The first gene bank collection of lentil was assembled in the Soviet Union by the Vavilov Institute of Plant Industry. The gene bank established in Turkey is rich with lentil landraces (Atikyilmaz, 2010), followed by Pakistan (Sultana and Ghafoor, 2008), Ethiopia (Fikiru et al., 2007), and China (Liu et al., 2008). At present, International Centre for Agricultural Research in Dry Areas (ICARDA) holds the 13,907 accessions of cultivated lentil and 603 accessions of wild lentil species (Singh and Chung, 2016). Wild lentil accessions represent <1% of global germplasm of this genus. The use of diverse germplasm in breeding lentil cultivars is still limited as only a few common lines have repeatedly been used in hybridization resulting in a narrow genetic base. In-situ and ex-situ conservation are the most efficient methods to conserve genetic diversity of lentil resources.

3.3.1 In-Situ Conservation

In-situ conservation means the conservation of a species in its natural habitat and the maintenance and recovery of viable population of species in their original place. It retains the material in its original location, where it was found, and it conserves the natural process of evolution, which is not possible in case of ex-situ conservation. Turkey is apparently the first country to produce a national plan for in-situ conservation of genetic diversity (Kaya et al., 1997). The genetic diversity of lentils' wild relatives is rich in the areas of Turkey and Syria. The distribution of all four wild taxa of genus *Lens* overlaps in the region of Aegean and the southwestern region (Ferguson et al., 1996). Further, an important area to target in-situ conservation includes West Turkey for *L. nigricans*; Northwest Syria, Southeast Turkey, South Syria, and Jordan for *L. culinaris* ssp. *orientalis*; coastal border region between Turkey and Syria stretching along the Syrian coast for *L. ervoides*; and South Syria for *L. culinaris* ssp. *odemensis* (Ferguson and Robertson, 1996). But, unfortunately, many areas of Turkey and other Mediterranean countries are threatened with the loss of invaluable genetic diversity (Solh and Erskine, 1981). However, *L. culinaris* ssp. *odemensis* and

L. ervoides are the wild lentil species which are most vulnerable to the loss of alleles (Ferguson et al., 1996). An exact status of on-farm, in-situ conservation of diverse germplasm is not well documented (Furman et al., 2009). The seed has been collected from each taxon and used in further study to determine the diversity within the population, which helps to establish the potential of in-situ conservation for wild *Lens* species (Ferguson and Robertson, 1996).

3.3.2 Ex-Situ Conservation

Ex-situ conservation means the conservation of species outside their natural habitats. It involves the transfer of genetic material away from the location where it is found. It ensures easy accessibility of germplasm for evaluation and utilization. Seeds of lentils are generally orthodox in nature, and can be dried to a lower moisture content of ±6% with minimum loss of viability. Seeds are stored at sub-zero temperature for maintaining viability for decades. Seeds are conserved for >50 years without any loss of viability (Usberti and Gomes, 1998). In ex-situ conservation of seeds, seeds can be stored as active (medium term) and base (long term) collections. For conservation of an active collection, viability must remain above 65% for 10–20 years, whereas in a base collection, seed are stored at −20°C to ensure long-term viability of seeds for >50 years.

The base collections are regularly monitored for seed viability at an interval of 10 years. Ex-situ conservation is the most convenient, cost effective, and widely used method of conservation. The periodic monitoring of the viability and timely regeneration of the materials is an essential part of ex-situ conservation, and vary according to the crop species and their reproductive systems (Breese, 1989). The Global Crop Diversity Trust (GCDT) supports the development of global crops (including some legumes) and regional strategies for ex-situ conservation and utilization of crop diversity. Approaches to ex-situ conservation of genetic resources include different methods like field gene banks, seed storage, and botanical gardens. DNA and pollen storage also contribute indirectly to ex-situ conservation of PGR (Fig. 3.1). The most widely used technique for legume species conservation is seed storage in gene banks, because legume seeds are generally known to be desiccation and freeze tolerant (Hong et al., 1998).

Globally, about 1.1 million germplasm accessions of grain legumes are conserved in various gene banks. ICARDA has the global mandate for research on improvement of the lentils and its conservation. The lentil germplasm accessions with ICARDA have been obtained from 113 global collection missions, and they include 1734 cultivated and 374 wild accessions. The major collections of crop species were carried out in the region of the "Fertile Crescent" in Western Asia, where the lentil was domesticated. Further, ICARDA gene banks have global responsibility for the conservation of the lentil germplasm belonging to 70 countries, 1146 ICARDA breeding lines, and 583 accessions of 6 wild lentil taxa, representing 23 countries (Alabboud et al., 2009). The whole conservation status of wild species in ICARDA gene bank is given in Table 3.1.

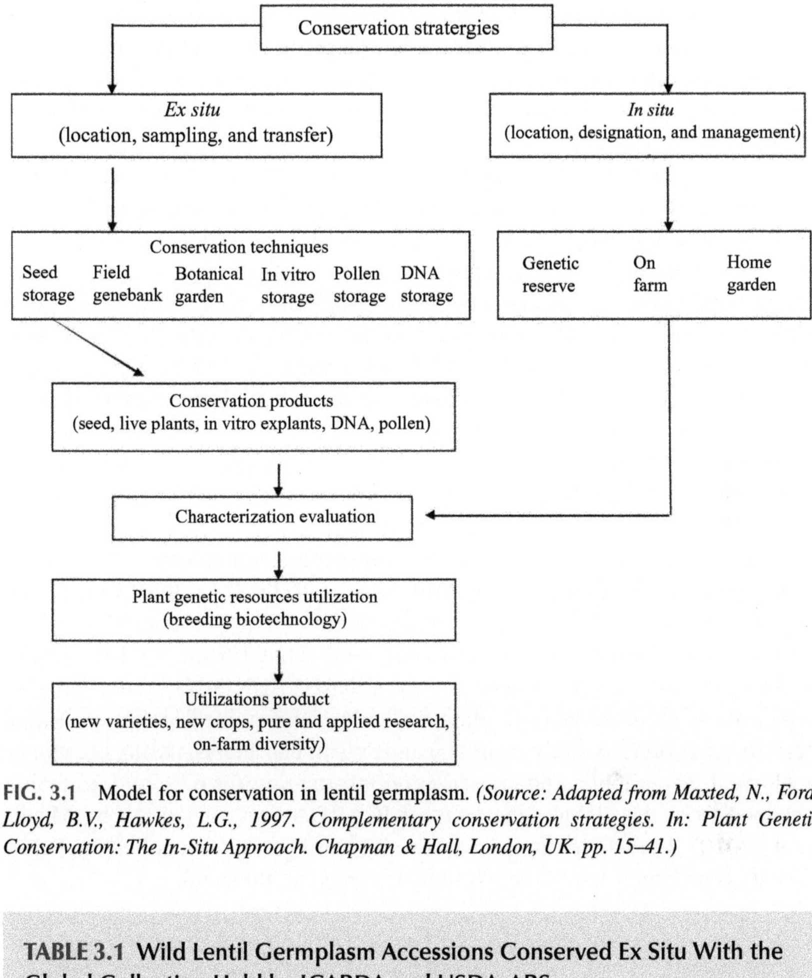

FIG. 3.1 Model for conservation in lentil germplasm. *(Source: Adapted from Maxted, N., Ford-Lloyd, B.V., Hawkes, L.G., 1997. Complementary conservation strategies. In: Plant Genetic Conservation: The In-Situ Approach. Chapman & Hall, London, UK. pp. 15–41.)*

TABLE 3.1 Wild Lentil Germplasm Accessions Conserved Ex Situ With the Global Collection Held by ICARDA and USDA-ARS

Species/Taxa	USDA	ICARDA
L. culinaris ssp. *orientalis*	92	268
L. culinaris ssp. *odemensis*	8	65
L. culinaris ssp. *tomentosus*	0	11
L. ervoides	61	166
L. lamottei	0	10
L. nigricans	37	63
Total	198	583

Source: Coyne, C., Rebecca, M., 2013. Lentil. In: Singh, M., Upadhyaya, H.D., Bisht, I.S. (Eds.), Genetic and Genomic Resources of Grain Legume Improvement. Elsevier, pp. 157–180.

Likewise, in India, National Bureau of Plant Genetic Resources (NBPGR) is recognized as an independent institute for the preservation of genetic resources over the past four decades, which has prioritized the collection, conservation, distribution, and documentation of the genetic materials of important crops, including lentils. Since its inception, systematic explorations have been conducted to augment lentil germplasm; area-specific and trait-specific collection trips were also undertaken. The major collection of lentils was undertaken from the central and eastern parts of Uttar Pradesh, Uttarakhand, the northern districts of Himachal Pradesh, Haryana, Rajasthan, and parts of Bihar (Malik et al., 2001). In the NATP projects, an effort for exploration and collection of lentil germplasm was intensified. The most recent inventory that includes 2655 accessions of cultivated lentil comprised of 2083 indigenous and 572 exotic accessions, which were introduced from different countries, have been conserved in the National Gene Bank of NBPGR (Table 3.2). The collected germplasm showed substantial variation in morpho-agronomic traits like plant growth habit, stem and foliage color, seedling color, high pod plant^{-1}, variation in seed coat color and grain size (Singh and Rana, 2006).

Furthermore, information can be retrieved on germplasm collections by Gene bank Information Management System (GIMS) at ICRISAT, GRIN-Global used at USDA-ARS (Cyr et al., 2009), PGR-Portal used at ICAR-NBPGR (http://pgrportal.nbpgr.ernet.in), and SINGER at CGIAR centers at the system-wide system. The European-based EURISCO system provides information about the ex-situ plant collections maintained in Europe (http://eurisco.ecpgr.org/). A study on in-situ and ex-situ gap analysis using taxonomic, ecological, geographic, and conservation information for 672 wild accessions obtained from the global gene bank of ICARDA was carried out by Maxted et al. (2010). The whole summary of ex-situ lentil germplasm collection held in various gene banks across the world is presented in Table 3.3.

TABLE 3.2 Lentil Germplasm Accessions Conserved Ex-Situ at NBPGR New Delhi

Crop	Indigenous	Exotic	Total
L. culinaris ssp. *culinaris*	2083	572	2655
L. culinaris ssp. *odemensis*	0	18	18
L. ervoides	0	48	48
L. ervoides	0	22	22

Source: Singh, J.R., Chung, G.H., 2016. Landmark research for pulses improvement. Indian J. Genet., 76, 399–409.

TABLE 3.3 The Global Ex-Situ Lentil Collection Held by the Global Gene Bank of ICARDA and Other National Gene Banks

Country	Gene Bank/Institute	Total Accessions	Wild Relatives	Land Races	% Collection Safety Duplicates
Syria	International Centre for Agricultural Research in Dry Areas	10,822	583	82%	
Australia	Australian Temperate Field Crops Collection	5254	04%	54%	67%
Iran	Seed and Plant Improvement Institute	3000	09%	12%	
United States	United States of Department of Agriculture	2875	01%	14%	94%
Russian Federation	N.I. Vavilov Russian Scientific Research Institute of Plant Industry	2556	11%	80%	100%
India	National Bureau of Plant Genetic Resources	2285	05%	42%	
Chile	Inst de Inv. Agropecuarias, Centro Regional de Investigación Carillanca	1345	10%	15%	
Canada	Plant Genetic Resource Centre	1139	17%	56%	
Turkey	Plant Genetic Resource Department Aegean Agricultural Research Institute	1095	01%	99%	
Syria	General Commission for Scientific Agricultural Research	1072	07%	38%	
Hungary	Research Centre for Agrobotany	1061	04%	03%	100%
Egypt	National Gene Bank	875	05%	5%	
China	Institute of Crop Germplasm Resources	855	10%	60%	100%

Continued

TABLE 3.3 The Global Ex-Situ Lentil Collection Held by the Global Gene Bank of ICARDA and Other National Gene Banks—cont'd

Country	Gene Bank/Institute	Total Accessions	Wild Relatives	Land Races	% Collection Safety Duplicates
Pakistan	Plant Genetic Resources Institute, National Agricultural Research Center	805	08%	91%	29%
Bangladesh	Bangladesh Agricultural Research Institute	798	Unknown	Unknown	
Spain	Centro de Recourses Fitogeneticos	703	10%	87%	
Ethiopia	Biodiversity Conservation and Research Institute	678	70%	Unknown	
Ukraine	Institute of Plant Production V.J. Yurjev of UAAS, Kharkiv	666	01%	52%	61%
Chile	Instituto de Investigaciones Agropecuarias, C.R.I. La Platina	600	Unknown	Unknown	
Israel	Agricultural Research Organization	500	Unknown	Unknown	
Nepal	Nepal Agricultural Research Council, Agricultural Botany Division	489	01%	97%	100%
Chile	Centro Regional de Investigación Quilamapu INIA	450	Unknown	64%	100%
Portugal	Estacao Nacional Melhoramento Plantas, Elvas	423	01%	02%	60%
Morocco	Institute National de la Recherche Agronomique INRA	365	Unknown	Unknown	
Bulgaria	Institute for Plant Genetic Resources "K.Malkov"	361	Unknown	16%	
Italy	Istituto di GeneticaVegetale (IGV)-Bari	348	01%	Unknown	
Spain	Banco de Germoplasma, Centro de Investigacion, Agraria de Albaladejito	321	01%	21%	

Ecuador	Instituto de Ciencias Naturales Universidad Central del Ecuador	295	03%	28%	
Ecuador	Estacion Experimental Santa Catalina, DENAREF, INIAP	252	01%	70%	13%
Slovakia	Research Institute of Plant Production Piestany	239	Unknown	Unknown	
Poland	Plant Breeding Station	216	Unknown	Unknown	
Mexico	Estación de Iguala, InstitutoNacional de Investigaciones Agrícolas	200	Unknown	Unknown	
Tunisia	Minister of Agriculture	65	Unknown	Unknown	
Yemen	Agricultural Research and Extension Authority	60	01%	14%	
Armenia	Institute of Botany, Botanical Gardens	34	Unknown	Unknown	
Tadjikistan	Makhsumov Scientific Reserch-Production Center "Ziroatkor" & Scientific Research Institute of Farming & TJK-Genebank (MSRPC)	28	Unknown	Unknown	
Azerbaijan	Agricultural Research Institute (ARI)	27	Unknown	Unknown	
Turkemenistan	Turkmen Scientific Research Institute for Cereals	22	Unknown	Unknown	
Azerbaijan	Azerbaijan National Academy of Sciences, Genetic Resources Institute	15	Unknown	Unknown	
Portugal	Banco de Germoplasma Genetica, EAN Oieras	15	1%	40%	100%
Portugal	Banco Portuges de Germplasm, Braga	05	Unknown	Unknown	

Source: http://www.croptrust.org.

3.3.3 Duplicates

The presence of duplicate accession is the major concern for the germplasm conservation in existing collections for all the gene banks worldwide. It is the prime responsibility of the curator to identify duplicate accessions and rationalize the gene bank collection. Different gene banks may use different techniques for the identification of duplicates and collection rationalization. ICAR-NBPGR developed a software R package "PGRdup" for removing duplicate accessions from the existing collection in gene bank by using passport information (Aravind et al., 2017).

3.3.4 Safety Duplicate

Duplication of genetically identical subsamples of accessions is required to mitigate the risk of its partial or total loss caused by natural or manmade catastrophes. The safety duplicates are genetically identical to the base collection and are referred to as the secondary most original sample (Enguls and Visser, 2003). Safety duplicates include both the duplication of material and its related information, and are deposited in the base collection at a different location, usually in another country. The location is chosen to minimize possible risks and provides the best possible storage facilities. About 7712 accessions of lentil species from ICARDA are maintained as safety duplicates in National gene bank of NBPGR at New Delhi. These duplicates are generally under a "black-box" approach; repository gene banks are not entitled to the use and distribution of the germplasm. It is the depositor's responsibility to ensure that the deposited material is of high quality, to monitor seed viability over time, and to use their own base collection to regenerate the collections when they begin to lose viability. The germplasm is not touched without permission from the depositor and is only returned on request when the original collection is lost or destroyed. Safety duplication is done for all original seeds collected by the gene bank or only held by the gene bank. Seeds which are duplicates from other collections can usually be retrieved from those collections and do not require safety duplication unless there is doubt about their security in other collections. There is a lack of duplicate collection and related information across the globe in case of lentil germplasm, which may lead to risk of loss.

3.4. GERMPLASM CHARACTERIZATION AND EVALUATION

The cultivated lentil has a narrow genetic base and low amount of molecular variation (Ferguson et al., 2000) which limits crop improvement progress. Several abiotic and biotic stresses also limit lentil production worldwide, while heat and drought are major ones affecting yield potential (Kumar et al., 2016a,b). Considerable genetic variability has been observed in lentil germplasm for agro-morphological and phenological traits (Erskine et al., 1989; Zaccardelli et al., 2012; Cristóbal et al., 2014). Characterization of 156 Ethiopian landrace populations representing 10 provinces of Ethiopia reveals geographically differentiated variation (Bejiga et al., 1996). Similarly, a wide range of variation was reported with respect to plant height, internode length, and 100-seed weight in a Turkish landrace (Toklu et al., 2009a,b). Characterization of cultivated lentil germplasm for some of the quality traits has been presented in Table 3.4.

Germplasm stocks derived from pure-line selection of landraces have also played a significant role as a source of novel alleles in conventional breeding programs. For example, Uthfala (Barimasur-1 = ILL 5888) was used as a potential donor for developing cultivars against Fusarium wilt and Ascochyta blight in Bangladesh (Sarker et al., 1999). Limited information exists on elite germplasm toward development of resistance against biotic stresses such as Stemphylium blight (Saha et al., 2010a), Fusarium vascular wilt (Eujayl et al., 1998; Hamwieh et al., 2005), Anthracnose (Tullu et al., 2003), Ascochyta blight (Ford et al.,

TABLE 3.4 List of Promising Germplasm Identified for Quality Traits

Traits	Importance	Promising Material	References
Green seed coat color	Higher market value	1294M-23 (a breeding line)	Davey (2007)
Tryptophan content	Synthesis of nicotinic acid and melatonin	Precoz (418.1 mg/100 g) and L 4163 (386.7 mg/100 g)	Solanki et al. (1999)
Protein	Structural, functional, and storage roles	Sinai 1 (31.52%), early maturity	Hamdi et al. (1996)
Starch	Functional and physicochemical properties	Richlea (45.3 g/100 g flour DM); ILL 5684 (44.2 g/100 g flour DM)	Tahir et al. (2011)
Minerals			
Fe	Involved in proper growth and development	L4704 with 125 ppm Fe (registered by NBPGR)	Sarker (2014)
Zn	Lowers absorption of intestinal fats	L4704 with 74 ppm Zn (registered by NBPGR)	Sarker (2014)
Antinutritional factors			
Tannin	Changes lentil color to brown/dark brown on cooking	P.I. 345,635 (Zero tannin gene)	Matus et al. (1993)
Raffinose family oligosaccharides (RFOs)	Degradation of RFOs causes diarrhea, discomfort, and flatulence	1156-2-17 (4.5 g/100 g flour DM); ILL 5684 (4.9 g/100 g flour DM)	Tahir (2011)

1999; Rubeena et al., 2006; Tullu et al., 2006), and lentil rust (Kant et al., 2004; Saha et al., 2010b). Similarly, no germplasm is available for a desirable degree of resistance to cold, which makes it possible to breed winter hardiness cultivars that can be grown well in highlands with a reasonable expectation of survival during winters (Erskine et al., 1981). Kahraman et al. (2004) also identified winter tolerant lines of *L. culinaris* ssp. *culinaris* ILL 662, ILL 857, ILL 975, and ILL 1878, which can be utilized as gene sources for developing winter tolerant cultivars. The foliar biotic stress Ascochyta blight incited by *A. lentis* is a serious lentil disease especially in Canada, Australia, Latin America, Ethiopia, Pakistan, the Northwest plains of India, United States, and highlands of WANA region (Johansen et al., 1994; Erskine et al., 1994; Ahmed and Morrall, 1996). Further, genetic gain is only possible by introgressing useful traits from the elite germplasm and wild species for widening the genetic base of cultivated varieties. It can be accomplished through precise germplasm characterization, which refers to the categorization of entire gene pool through distinct morphological traits. These traits can help to distinguish distinct phenotypic groups, and their inheritance patterns can be studied using principles of classical genetics. The phenotypic evaluation permits identification of specific alleles for specific gene loci (Muehlbauer and Singh, 1987; Pundir et al., 1985). Ferguson et al. (1998a) characterized a set of 310 accessions of wild lentil species and found substantial range of variation for more number of leaves, peduncle, and pods per plant. Gupta and Sharma (2006) evaluated a set of 93 accessions of wild lentil species and found sources of resistance for rust, powdery mildew, and Fusarium wilt. They also reported that quaternary gene pool of lentil where *L. nigricans* provided the maximum no. of accessions for biotic and abiotic stress tolerance. Toklu et al. (2009a,b) characterized Turkish lentil landraces for important agro-morphological traits and found a wide range of diversity with respect to plant height, internode length, and 100-seed weight. Singh et al. (2014a,b) also characterized and evaluated 405 global wild *Lens* collections from the ICARDA gene bank under two agro-ecological regions of Northwestern India. The study showed a wide range of intraspecific and interspecific variation for the majority of morphological plant characteristics. Seedling stem pigmentation exhibited variation in all the *Lens* taxa except *L. culinaris* ssp. *culinaris*. Likewise, leaf pubescence, leaflet size, and tendril length also revealed remarkable variation in all wild *Lens* taxa. There was a substantial variation in pod shedding and dehiscence for a majority of the *Lens* species except *L. culinaris* ssp. *culinaris*. Flower color was white and purple in majority of wild species, except *L. ervoides,* where the color appeared to be purple only. The ground color of testa was mostly gray and brown in all the species except, *L. culinaris* ssp. *culinaris*, where it was all brown. Substantial variation was also found in the pattern of testa for all the wild species. Majority of wild accessions had yellow and orange cotyledon color except *L. culinaris* ssp. *tomentosus*, where it was orange. The summary of evaluation of wild species against important agro-morphological traits including major biotic and abiotic stresses is presented in Table 3.5.

TABLE 3.5 Sources of Useful Traits Identified in Lentil Landraces and Their Wild Relatives

Trait of Interest	Wild Relative(s)	Reference(s)
Biotic stress		
Powdery mildew resistance	*L. culinaris* ssp. *orientalis, L. culinaris* ssp. *odemensis, L. ervoides, L. nigricans, L. culinaris* ssp. *tomentosus*	Gupta and Sharma (2006), Singh et al. (2014a, b), and Gautam et al. (2013)
Resistance to Orobanche	*L. ervoides, L. culinaris* ssp. *odemensis, L. culinaris* ssp. *orientalis*	Femandez-Aparicio et al. (2008) and Femandez-Aparicio et al. (2009)
Rust resistance	*L. culinaris* ssp. *orientalis, L. culinaris* ssp. *odemensis, L. ervoides, L. nigricans, L. culinaris* ssp. *orientalis, L. lamottei*	Gupta and Sharma (2006), Singh et al. (2014a, b), and Gautam et al. (2013)
Resistance to sitona weevils	*L. culinaris* ssp. *odemensis, L. ervoides, L. nigricans, L. culinaris* ssp. *orientalis*	Bouhssini et al. (2008)
Vascular wilt resistant	*L. culinaris* ssp. *orientalis, L. culinaris* ssp. *odemensis, L. ervoides, L. nigricans*	Gupta and Sharma (2006) and Bayaa et al. (1995)
Anthracnose resistance	*L. ervoides, L. lamottei, L. nigricans*	Tullu et al. (2006), Tullu et al. (2010), Fiala et al. (2009), Vail and Vandenberg (2011), and Bhadauria et al. (2017)
Resistance to bruchid weevils	*L. culinaris* ssp. *orientalis, L. nigricans, L. lamottei*	Laserna-Ruiz et al. (2012)
Fusarium wilt resistance	*L. culinaris* ssp. *orientalis, L. ervoides*	Bayaa et al. (1997), Gupta and Sharma (2006), and Mohammadi et al. (2011)
Callosobruchis chinensis	*L. ervoides*	Gore et al. (2015)
Ascochyta blight resistance	*L. ervoides, L. culinaris* ssp. *orientalis, L. culinaris* ssp. *odemensis, L. lamottei, L. nigricans*	Bayaa et al. (1994), Tullu et al. (2006, 2010), Nguyen et al. (2001), and Gautam et al. (2013)
Root-knot nematode	*L. culinaris* ssp. *odemensis, L. nigricans*	Gautam et al. (2013)
Stemphylium blight	*L. ervoides, L. culinaris* ssp. *orientalis, L. tomentosus, L. nigricans, L. culinaris* ssp. *odemensis, L. lamottei*	Podder et al. (2013) and Bhadauria et al. (2017)

Continued

TABLE 3.5 Sources of Useful Traits Identified in Lentil Landraces and Their Wild Relatives—cont'd

Trait of Interest	Wild Relative(s)	Reference(s)
Colletotrichum truncatum resistance	*L. culinaris* ssp. culinaris, *L. ervoides*	Buchwaldt et al. (2004), Shaikh et al. (2012), Fiala et al. (2009), and Bhadauria et al. (2017)
Abiotic stress		
Cold tolerance	*L. culinaris* ssp. *orientalis*	Hamdi and Eriskine (1996)
Drought tolerance	*L. culinaris* ssp. *odemensis*, *L. ervoides*, *L. nigricans*	Gupta and Sharma (2006) and Hamdi and Eriskine (1996)
Heat tolerance	*L. culinaris* ssp. *culinaris*	Kumar et al. (2016a, b)
Winter hardiness	*L. culinaris* ssp. *orientalis*	Hamdi et al. (1996)
Agronomic characters		
Yield attributes	*L. culinaris* ssp. *orientalis*, *L. culinaris* ssp. *odemensis*	Gupta and Sharma (2006) and Singh et al. (2014a, b)
Seed yield per plant	*L. ervoides*	Gautam et al. (2013)
Biofortification		
Minerals and high protein content	*L. culinaris* ssp. *orientalis*, *L. culinaris* ssp. *odemensis*, *L. ervoides*, *L. nigricans*, *L. culinaris* ssp. *tomentosus*, *L. ervoides*	Kumar et al. (2016a, b) and Kumar et al. (2018)

3.5. DEVELOPMENT OF CORE COLLECTIONS

The core collection concept was developed in the 1980s (Frankel and Brown, 1984) for effective management of genetic resources and to make germplasm more accessible for potential users. It represents about 10% of the total entire collection that captures the maximum variability prevalent in the entire sample size, but depending on the diversity and size of the collection, that number may vary from 5% to 20%. The core collection should serve as a working collection and should be extensively examined, and the accessions, which are not included in the core collection, should be designated as reserve collection. The large size of collections and lack of reliable data on traits of economic importance shows high genotype × environment interactions, which is considered the main reason for low use of genetic resources. ICARDA has developed 1000 accessions of composite collection, which represents genetic diversity of all global collection (Furman, 2006). The core collection comprised of 287 accessions of lentil, which were selected based on country of origin at Regional Plant Introduction Station (USDA) located at Pullman, WA, United States, has been characterized and evaluated for morphology, phenology, biomass, and seed yield over two seasons (Tullu et al., 2001). Singh et al. (2014a,b) also developed core set of 96 accessions from the global wild lentil species. These core collections were further validated and evaluated for various traits of interest.

Maintenance of germplasm for any species depends on its reproductive system. Lentils, as a self-pollinated crop, have natural outcrossing of $<0.8\%$ (Wilson and Law, 1972). The rate of outcrossing depends upon several factors such as cultivars, year, and location, which differ within the cultivars (Horneburg, 2006). Globally, gene bank of ICARDA ensures the maintenance of improved germplasm of lentils, per international standards. Maintenance comprises seed viability, seed vigor, and seed quantity of lentil accessions in the gene bank. The genetic integrity, quality, and quantity of germplasm can be maintained by regular intervals of monitoring the health of conserved seeds in the gene bank. Regeneration is one of the most crucial steps in the gene bank activities which is carried out in the field without any control on pollination due to lentil cleistogamy. Top priorities for regeneration are given to those accessions which have poor quality (low viability) and lower quantity of seed. Sackville Hamilton and Chorlton (1997) published the guidelines for the regeneration of lentil germplasm.

3.6. CONCLUSION AND FUTURE PROSPECTS

The lentil taxa, including wild relatives, are a potential resource of various untapped genes controlling desirable traits for high yield, and resistance to biotic and abiotic stresses, to make substantial improvements in productivity, thereby widening the genetic base of elite lentil germplasm. Desired genes from wild lentils belonging to different gene pools have been successfully transferred into the cultivated lentil background using conventional and biotechnological approaches. However, to harness the full potential of useful genes from tertiary gene pools, efficient tissue culture protocols, advances in genetic engineering, and development of strategies to avoid linkage drag would facilitate the free flow of genes. Molecular markers could be used as valuable tools in every breeding program for their role in identification of donor parents, true F_1 hybrids, and backcrosses to efficient prebreeding and phylogenetic analysis. The studies pertaining to climate resilience will also ensure effective conservation of lentil germplasm.

REFERENCES

Ahmed, S., Morrall, R.A.A., 1996. Field reactions of lentil lines and cultivars to isolates of *Ascochyta fabae* f.sp. *lentis* on lentil. Can. J. Plant Pathol. 18, 362–369.

Alabboud, A., Szilagyi, L., Roman, G.H.V., 2009. Assessment of genetic diversity in lentil (*Lens culinaris* Medik.) as revealed by RAPD markers. In: USAMV Bucharest Series A. vol. LII, pp. 439–444.

Aravind, J., Radhamani, J., Srinivasan, K., Subhash, B.A., Tyagi, R.K., Aubrey, M., Atkinson, K., Philips, L., 2017. Package 'PGRdup' Title Discover Probable Duplicates in Plant Genetic Resources Collections. Available from: https://cran.r-project.org/package=PGRdup.

Arumuganathan, K., Earle, E.D., 1991. Nuclear DNA content of some important plant species. Plant Mol. Biol. Rep. 9, 208–218.

Atikyilmaz, N., 2010. In: Grain legumes genetic resources activities in Turkey. ECPGR Working Group on Grain Legumes, 4th Meeting, 16–17 November 2007, Lisbon.

Bahl, P.N., Lal, S., Sharma, B.M., 1993. An overview of the production and problems in Southeast Asia. In: Erskine, W., Saxena, M.C. (Eds.), Lentil in South Asia. Proceedings of the Seminar on Lentils in South Asia. ICARDA, Aleppo, Syria, pp. 1–10.

Bayaa, B., Erskine, W., Hamdi, A., 1994. Geographic distribution of resistance to Ascochyta blight in wild lentil. Genet. Resour. Crop. Evol. 41, 61–65.

Bayaa, B., Erskine, W., Hamdi, A., 1995. Evaluation of a wild lentil collection for resistance to vascular wilt. Genet. Resour. Crop. Evol. 42, 231–235.

Bayaa, B., Erskine, W., Singh, M., 1997. Screening lentil for resistance to *Fusarium* wilt: methodology and sources of resistance. Euphytica 98, 69–74.

Bejiga, G., Tsegaye, S., Tullu, A., Erskine, W., 1996. Quantitave evaluation of Ethiopian landraces of lentil (*Lens culinaris*). Genet. Resour. Crop. Evol. 43, 293–301.

Bhadauria, V., Ramsay, L., Bett, K.E., Banniza, S., 2017. QTL mapping reveals genetic determinants of fungal disease resistance in the wild lentil species *Lens ervoides*. Sci. Rep. 7. https://doi.org/10.1038/s41598-017-03463-9.

Bhatty, R.S., 1988. Composition and quality of lentil: a review. Can. Inst. Food Sci. Technol. J. 21, 144–160.

Bouhssini, E.I., Sarker, M., Eriskine, W., Joubi, A., 2008. First sources of resistance to Sitona weevil (*Sitona crinitus* Herbst) in wild *Lens* species. Genet. Resour. Crop. Evol. 55, 1–4.

Breese, E.L., 1989. Regeneration and Multiplication of Germplasm Resources in Seed Gene Banks: The Scientific Background. International Board of Plant Genetic Resources, Rome.

Buchwaldt, L., Anderson, K.L., Morrall, R.A.A., Gossen, B.D., Bernier, C.C., 2004. Identification of lentil germplasm resistant to *Colletotrichum truncatum* and characterization of two pathogen races. Phytopathology 94, 236–243.

Cristóbal, M.D., Pando, V., Herrero, B., 2014. Morphological characterization of lentil (*Lens culinaris* Medik.) landraces from Castilla Y León, Spain. Pak. J. Bot. 46, 1373–1380.

Cubero, J.I., 1981. Origin, domestication and evolution. In: Webb, C., Hawtin, G.C. (Eds.), Lentils. Commonwealth Agricultural Bureau, Slough, UK, pp. 15–38.

Cyr, P.D., Weaver, B., Millard, M.J., Gardner, C.A., Bohning, M.A., Emberland, G.P., Sinnott, Q.P., Kinard, G.R., Franco, T., Mackay, M., Guarino, L., Bretting, P.K., 2009. In: GRIN-Global: an international project to develop a global plant gene bank and information management system. Plant & Animal Genome XVII: International Conference on the Status of Plant & Animal Genome Research, Poster Abstract Number 800, pp. 136.

Davey, B.F., 2007. Green Seed Coat Colour Retention in Lentil (*Lens culinaris*). (Doctoral dissertation).

Enguls, J., Visser, L. (Eds.), 2003. A Guide of Effective Management of Germplasm Collection. Bioversity International. (no. 6).

Eriskine, W., Rihawe, S., Capper, B.S., 1990. Variation in lentil straw quality. Anim. Food Sci. Technol. 28, 61–69.

Erskine, W., Sarker, A., Kumar, S., 2011. Crops that feed the world 3. Investing in lentil improvement toward a food secure world. Food Sec. 3, 127–139.

Erskine, W., Witcombe, J.R., 1984. Lentil gemplazm catalog. International Center for Agricultural Research in the Dry Areas (ICARDA), Aleppo, Syria.

Erskine, W., Myveci, K., Izgin, N., 1981. Screening a world lentil collection for cold tolerance. Lens Newsl. 8, 5–8.

Erskine, W., Adham, Y., Holly, L., 1989. Geographic distribution of variation in quantitative traits in a world lentil collection. Euphytica 43, 97–103.

Erskine, W., Tufail, M., Russell, A., Tyagi, M.C., Rahman, M.M., Saxena, M.C., 1994. Current and future strategies in breeding lentil for resistance to biotic and abiotic stresses. Euphytica 73, 127–135.

Eujayl, I., Erskine, W., Bayaa, B., Baum, M., Pehu, E., 1998. *Fusarium* vascular wilt in lentil: inheritance and identification of DNA markers for resistance. Plant Breed. 117, 497–499.

FAO, 2010. The Second Report on the State of the world's Plant Genetic Resources for Food and Agriculture, Rome.

FAO, 2015. FAOSTAT, Food and Agriculture Organization of the United Nations, Rome. Available from: http://faostat.fao.org.

Femandez-Aparicio, M., Sillero, J.C., Rubiales, D., 2008. Identification of sources of resistance to crenate broomrape (*Orobanche crenata*) in Spanish lentil (*Lens culinaris*) germplasm. Weed Res. 48, 85–94.

Femandez-Aparicio, M., Sillero, J.C., Rubiales, D., 2009. Resistance to broomrape in wild lentils (*Lens*spp.). Plant Breed. 128, 266–277.

Ferguson, M.E., Robertson, L.D., 1996. Genetic diversity and taxonomic relationships within the genus *Lens* as revealed by allozyme polymorphism. Euphytica 91, 163–172.

Ferguson, M., Acikgoz, N., Ismail, A., Cinsoy, A., 1996. An eco-geographic survey of wild *Lens* species in Aegean and south West Turkey. Anadolu 6, 159–166.

Ferguson, M.E., Robertson, L.D., Ford-Lloyd, B.V., Newbury, H.J., Maxted, N., 1998a. Contrasting genetic variation amongst lentil landraces from different geographical origins. Euphytica 102, 265–273.

Ferguson, M.E., Ford-Llyod, B., Robertson, L.D., Maxted, N., Newbury, H.J., 1998b. Mapping the geographical distribution of genetic variation in the genus *Lens* for the enhanced conservation of plant genetic diversity. Mol. Ecol. 7, 1743–1755.

Ferguson, M.E., Maxted, N., Slageren, M.V., Robertson, L.D., 2000. A re-assessment of the taxonomy of *Lens* mill. (Leguminosae, Papilionoideae, Vicieae). Bot. J. Linn. Soc. 133, 41–59.

Fiala, J.V., Tullu, A., Banniza, S., Swartz, G.S., Vandenberg, A., 2009. Interspecies transfer of resistance to anthracnose in lentil (*Lens culinaris* Medik.). Crop Sci. 49, 825–830.

Fikiru, E., Tesfaye, K., Bekele, E., 2007. Genetic diversity and population structure of Ethopian lentil (*Lens culinaris* Medikus) landraces as revealed by ISSR marker. Afr. J. Biotechnol. 6, 1460–1468.

Ford, R., Pang, E.C.K., Taylor, P.W.J., 1999. Genetics of resistance to Ascochyta blight (*A. lentis*) of lentil and the identification of closely linked RAPD markers. Theor. Appl. Genet. 98, 93–98.

Frankel, O.H., Brown, A.H.D., 1984. Plant genetic resources today: a critical appraisal. In: Holden, J.H.W., Williams, J.T. (Eds.), Crop Genetic Resources: Conservation & Evaluation. George Alien & Unwin Ltd., London, pp. 249–257.

Furman, B.J., 2006. Methodology to establish a composite collection: case study in lentil. Plant Genet. Resour. 4, 2–12.

Furman, B.J., Coyne, C., Redden, B., Sharma, S.K., Vishnyakova, M., 2009. Genetic resources: collection, characterization, conservation and documentation. In: Erskine, W., Muelbauer, F.J., Sarker, A., Sharma, B. (Eds.), The Lentil: Botany, Production and Uses. CABI International, Wallingford, Oxfordshire, UK, pp. 64–75.

Gautam, N.K., Singh, M., Khan, Z., Roy, A., Akhtar, J., Ram, B., 2013. Assessment of lentil germplasm with respect to agromonic performance and major biotic stress. Legum. Res. 36, 214–219.

Gore, P.G., Chauhan, S.K., Tripathi, K., Bhalla, S., 2015. Physical basis of resistance in wild *Lens* spp. accessions to pulse beetle, *Callosobruchus chinensis*. Indian J. Plant Protect. 43, 267–272.

Gupta, D., Sharma, S.K., 2006. Evaluation of wild *Lens taxa* for agro-morphological traits, fungal diseases and moisture stress in north western Indian Hills. Genet. Resour. Crop. Evol. 53, 1233–1241.

Hamdi, A., Eriskine, W., 1996. Reaction of wild species of the genus *Lens* to drought. Euphytica 91, 173–179.

Hamdi, A., Ktismenoglu, I., Erskine, W., 1996. Sources of winter hardiness in wild lentil. Genet. Resour. Crop. Evol. 43, 63–67.

Hamwieh, A., Udapa, S.M., Choumane, W., Sarker, A., Dreyer, F., Jung, C., Baum, M., 2005. A genetic linkage map of lentil based on microsatellite and AFLP markers and localization of fusarium vascular wilt resistance. Theor. Appl. Genet. 110, 669–677.

Hanelt, P., 2001. Lens Mill. In: Hanelt, P. (Ed.), Mansfeld's Encyclopedia of Agricultural and Horticultural Crops. vol. 2, pp. 849–852.

Harlan, J.R., 1992. Crops and Man, second ed. American Society of Agronomy, Madison, Wisconsin, USA284.

Hong, T.D., Linington, S., Ellis, R.H., 1998. Compendium of Information on Seed Storage Behaviour. UK Royal Botanic Gardens.

Horneburg, B., 2006. Outcrossing in lentil (*Lens culinaris*) depends on cultivar, location and year, and varies within cultivars. Plant Breed. 125, 638–640.

Johansen, C., Baldev, B., Brouwer, J.B., Erskine, W.K., Jermyn, W.A., Li-Juan, L., Malik, B.A., Miah, A.A., Silim, S.N., 1994. Biotic and abiotic stresses constraining productivity of cool season food legumes in Asia, Africa and Oceania. In: Muehlbaur, F.J., Kaiser, K.J. (Eds.), Expanding the production and Use of Cool Season Food Legumes. vol. 19. Current Plant Science and Biotechnology in Agriculture, Kluwer Academic, Dordrecht, Netherlands, pp. 175–194.

Kahraman, A., Kusmenoglu, I., Aydin, N., Aydogan, A., Erskine, W., Muehlbauer, F.J., 2004. QTL mapping of winter hardiness genes in lentil. Crop Sci. 44, 13–22.

Kant, A., Sharma, S.K., Sharma, R.K., Mohapatra, T., 2004. Identification of RAPD and AFLP markers linked with rust resistance gene in lentil. Crop Improvement India 31, 1.

Kaya, Z., Kun, E., Gruner, A., 1997. National Plan for In Situ Conservation of Plant Genetic Diversity in Turkey. Ministry of Environment, Ankara.

Kumar, S., Rajendran, K., Kumar, J., Hamweih, A., Baum, M., 2015. Current knowledge in lentil genomics and its application for crop improvement. Front. Plant Sci. 6, 78.

Kumar, J., Gupta, S., Gupta, P., Dubey, S., Singh, R.S., Kumar, S., 2016a. Breeding strategies to improve lentil for diverse agro-ecological environments. Indian J. Genet. 76, 530–549.

Kumar, J., Singh, J., Kanaujia, R., Gupta, S., 2016b. Protein content in wild and cultivated taxa of lentil (*Lens culinaris* ssp. *culinaris* Medikus). Indian J. Genet. 76, 631–634.

Kumar, S., Choudhary, A.K., Rana, K.S., Sarker, A., Singh, M., 2018. Bio-fortification potential of global wild annual lentil core collection. PLoS ONE 13, e0191122.

Ladizinsky, G., 1993. Wild lentils. Crit. Rev. Plant Sci. 12, 169–184.

Ladizinsky, G., 1997. A new species of *Lens* from south-East Turkey. Bot. J. Linn. Soc. 123, 257–260.

Laserna-Ruiz, I., De-Los-Mozos-Pascual, M., Santana-Meridas, O., Sanchez-Vioque, R., Rodriguez-Conde, M.F., 2012. Screening and selection of lentil (*Lens* Miller) germplasm resistant to seed bruchids (*Bruchus* spp.). Euphytica 73, 1–10.

Liu, J., Guan, J.P., Xu, D.X., Zhang, X.Y., Gu, J., Zong, X.X., 2008. Genetic diversity and population structure in lentil (*Lens culinaris* Medik) germplasm detected by SSR markers. Acta Agron. Sin. 34, 1901–1909.

Ljustina, M., Miki'c, A., 2010. Archaeological evidence for the domestication of lentil (*Lens culinaris*) and its distribution in Europe. J. Lentil Res. 4, 26–29.

Malik, S.S., Srivastava, U., Tomar, J.B., Bhandari, D.C., Pandey, A., Hore, D.K., Dikshit, N., 2001. Plant germplasm and germplasm collection. In: Dhillon, B.S., Varaprasad, K.S., Srinivasan, K., Singh, M., Archak, S., Srivastava, U., Sharma, G.D. (Eds.), A Compendium of Achievements. NBPGR, New Delhi, pp. 31–68.

Matus, A., Slinkard, A.E., Vandenberg, A., 1993. The potential of zero tannin lentil. New Crops 279–282.

Maxted, N., Hargreaves, S., Kell, S.P., Amari, A., Street, K., Shehadeh, A., Pigin, J., Konopka, J., 2010. In: Temperature forage and pulse legume genetic map analysis. XIII OPTIMA Meeting. Antalya, Turkey, pp. 22–26.

Mohammadi, N., Mohammadi Goltapeh, E., Babaie Ahari, A., Puralibaba, H., 2011. Pathogenic and genetic characterization of Iranian isolates of *Fusarium oxysporum* f.sp. *lentis* by ISSR analysis. J. Agric. Technol. 7, 63–72.

Muehlbauer, F.J., 1991. Use of introduced germplasm in cool season food legume cultivar development. In: Shands, H.L., Wiesner, L.E. (Eds.), Use of Plant Introductions in Cultivar Development (Part 2). Crop Sci. Soc. Amer. Social publication no. 20, pp. 49–73.

Muehlbauer, F.J., Singh, K.B., 1987. Genetics of chickpea. In: Saxena, M.C., Singh, K.B. (Eds.), The Chickpea. CABI International, Cambridge, pp. 99–126.

Nguyen, T.T., Tailor, P.W.J., Brouwer, J.B., Pang, E.C.K., Ford, R., 2001. A novel source of resistance in lentil (*Lens culinaris* ssp. *culinaris*) to *Ascochyta* blight caused by *Ascochyta lentis*. Aust. J. Plant Pathol. 30, 211–215.

Podder, R., Banniza, S., Vandenberg, A., 2013. Screening of wild and cultivated lentil germplasm for resistance to stem phylium blight. Plant Genet. Resour. 11, 26–35.

Pratap, A., Kumar, J., Kumar, S., 2014. Evaluation of wild species of lentil for agro-morphological traits. Legume Res. 37, 11–18.

Pundir, R.P.S., Rao, N.K., van der Maesen, L.J.G., 1985. Distribution of qualitative traits in the world germplasm of chickpea. Euphytica 34, 697–703.

Rehman, S., Altaf, C.H.M., 1994. Karyotipic studies in *Lens culinaris* Medic, ssp. Macrosperma cv. Laird X Precoz. Pak. J. Bot. 26, 347–352.

Rubeena, Taylor, P.W.J., Ades, P.K., Ford, R., 2006. QTL mapping of resistance in lentil (*Lens culinaris* ssp. *culinaris*) to ascochyta blight (*Ascochyta lentis*). Plant Breed. 125, 506–512.

Sackville Hamilton, S.N., Chorlton, K.H., 1997. Regeneration of Accessions in Seed Collections: A Decision Guide. Bioversity International. (no. 5).

Saha, G.C., Sarker, A., Chen, W.D., Vandemark, G.J., Muehlbauer, F.J., 2010a. Inheritance and linkage map positions of genes conferring resistance to stemphylium blight in lentil. Crop Sci. 50, 1831–1839.

Saha, G.C., Sarker, A., Chen, W.D., Vandemark, G.J., Muehlbauer, F.J., 2010b. Identification of markers associated with genes for rust resistance in *Lens culinaris* Medik. Euphytica 175, 261–265.

Sarker, A., 2014. Iron and zinc lentils. In: The Second Global Conference on Biofortification: Getting Nutritious Foods to People, conference brief #5b, Kigali, Rwanda.

Sarker, A., Erskine, W., Sharma, B., Tyagi, M.C., 1999. Inheritance and linkage relationship of days to flower and morphological loci in lentil (*Lens culinaris* Medikus subsp. *culinaris*. J. Hered. 90, 270–275.

Sehirali, S., 1988. Grain Legume Crops. Ankara University, Faculty of Agricultural Engineering, Ankara, Turkey435.

Shaikh, R., Diederichsen, A., Harrington, M., Adam, J., Conner, R.L., Buchwaldt, L., 2012. New sources of resistance to *Colletotrichum truncatum* race Ct0 and Ct1 in *Lens culinaris* Medikus subsp. *culinaris* obtained by single plant selection in germplasm accessions. Genet. Resour. Crop. Evol. 60, 193–201.

Sharpe, A.G., Ramsay, L., Sanderson, L.A., Fedoruk, M.J., Clarke, W.E., Rong, L., et al., 2013. Ancient orphan crop joins modern era: gene based-SNP discovery and mapping in lentil. BMC Genom. 14, 192.

Singh, J.R., Chung, G.H., 2016. Landmark research for pulses improvement. Indian J. Genet. 76, 399–409.

Singh, B., Rana, R.S., 2006. Pulses germplasm collection and enhancement at NBPGR: present accomplishments and future prospects. In: Asthana, A.N., Masood, A. (Eds.), Recent Advances in Pulses Research. Kanpur: Indian Society of Pulses Research and Development. Indian Institute of Pulses Research, pp. 743–755.

Singh, M., Bisht, I.S., Kumar, S., Dutta, M., Bansal, K.C., Karale, M., Sarker, A., Amri, A., Kumar, S., Dutta, S.K., 2014a. Global wild annual *Lens* collection: a potential resource for lentil genetic base broadening and yield enhancement. PLoS ONE 9, e107781.

Singh, M., Bisht, I.S., Dutta, M., Kumar, K., Kumar, S., Bansal, K.C., 2014b. Genetic studies on morpho-phenological traits in lentil (*Lens culinaris* Medikus) wide crosses. J. Genet. 93, 561–566.

Solanki, I.S., Kapoor, A.C., Singh, U., 1999. Nutritional parameters and yield evaluation of newly developed genotypes of lentil (*Lens culinaris* Medik.). Plant Foods Human Nutr. 54, 79–87.

Solh, M., Erskine, W., 1981. Genetic resources. In: Webb, C., Hawtin, G.C. (Eds.), The Lentil. Commonwealth Agriculture Bureau, Slough, UK, pp. 54–67.

Sonnante, G., Hammer, K., Pignone, D., 2009. From the cradle of agriculture a handful of lentils: history of domestication. Rend Lincei 20, 21–37.

Sultana, T., Ghafoor, A., 2008. Genetic diversity in ex-situ conserved *Lens culinaris* for botanical descriptors, biochemical and molecular markers and identification of landraces from indigenous genetic resources of Pakistan. J. Integr. Plant Biol. 50, 484–490.

Tahir, M., 2011. Characterization of Raffinose Family Oligosaccharides in Lentil Seeds. (Doctoral dissertation).

Tahir, M., Lindeboom, N., Baga, M., Vandenberg, A., Chibbar, R., 2011. Composition and correlation between major seed constituents in selected lentil (*Lens culinaris* Medik) genotypes. Can. J. Plant Sci. 91, 825–835.

Toklu, F., Karakoy, T., Hakli, E., Bicer, T., Brandolini, A., Kilian, B., Ozkhan, H., 2009a. Genetic variation among lentil (*Lens culinaris* Medik.) landraces from Southeast Turkey. Plant Breed. 128, 178–186.

Toklu, F., Tuba, B., Karakoy, T., 2009b. Agro-morphological characterization of the Turkish lentil landraces. Afr. J. Biotechnol. 8, 4121–4127.

Tullu, A., Kushmenoglu, I., McPhee, K.E., Muehlbauer, F.J., 2001. Characterization of core collection of lentil germplasm for phenology, morphology, seed and straw yields. Genet. Resour. Crop Evol. 48, 143–162.

Tullu, A., Buchwaldt, L., Warkentin, T., Taran, B., Vandenberg, A., 2003. Genetics of resistance to anthracnose and identification of AFLP and RAPD markers linked to the resistance gene in PI 320937 germplasm of lentil (*Lens culinaris* Medikus). Theor. Appl. Genet. 106, 428–434.

Tullu, A., Taran, B., Breitkreutz, C., Buchwaidt, L., Banniza, S., Warkentin, T.D., et al., 2006. A quantitative-trait locus for resistance to ascochyta blight *Ascochyta lentis* maps close to a gene for resistance to anthracnose *Colletotrichum truncatum* in lentil. Can. J. Plant Pathol. 28, 588–595.

Tullu, A., Banniza, S., Tar'an, B., Warkentin, T., Vandenberg, A., 2010. Sources of resistance to Ascochyta blight in wild species of lentil (*Lens culinaris* Medik.). Genet. Resour. Crop. Evol. 57, 1053–1063.

Usberti, R., Gomes, R.B.R., 1998. Seed viability constants for groundnut. Ann. Bot. 82, 691–694.

Vail, S., Vandenberg, A., 2011. Genetic control of inter-specific derived and juvenile resistance in lentil to *Colletotrichum truncatum*. Crop Sci. 51, 1481–1490.

Van Oss, H., Aron, Y., Ladizinsky, G., 1997. Chloroplast DNA variation and evolution in the genus *Lens* Mill. Theor. Appl. Genet. 94, 452–457.

Wilson, V.E., Law, A.G., 1972. Natural crossing in *Lens esculenta* Moench. Am. Soc. Horticult. Sci. J. 97, 142–143.

Zaccardelli, M., Lupo, F., Piergiovanni, A.R., Laghetti, G., Sonnante, G., Daminati, M.G., et al., 2012. Characterization of Italian lentil (*Lens culinaris* Medik.) germplasm by agronomic traits, biochemical and molecular markers. Genet. Resour. Crop. Evol. 59, 727–738.

Zohary, D., 1999. Monophyletic vs. polyphyletic origin of the crops on which agriculture was founded in the Near East. Genet. Resour. Crop Evol. 46, 133–142.

FURTHER READING

Coyne, C., Rebecca, M., 2013. Lentil. In: Singh, M., Upadhyaya, H.D., Bisht, I.S. (Eds.), Genetic and Genomic Resources of Grain Legume Improvement. Elsevier, pp. 157–180.

Maxted, N., Ford-Lloyd, B.V., Hawkes, L.G., 1997. Complementary conservation strategies. In: Plant Genetic Conservation: The In-Situ Approach. Chapman & Hall, London, UK, pp. 15–41.

Simon, C.J., Hannan, R.M., 1995. Development and use of core subsets of cool season food legume germplasm collection. Horticult. Sci. 30, 907.

Singh, M., Rana, M.K., Kumar, K., Bisht, I.S., Dutta, M., Gautam, N.K., Sarker, A., Bansal, K.C., 2013. Broadening the genetic base of lentil cultivars through inter-sub-specific and inter-specific crosses of *Lens* taxa. Plant Breed. 132, 667–675.

Chapter 4

Conventional Genetic Manipulations

Rakesh Kumar Chahota*, Tilak Raj Sharma†, Shyam Kumar Sharma‡
*Department of Agricultural Biotechnology, CSK Himachal Pradesh Agriculture University, Palampur, India
†ICAR-Indian Institute of Agricultural Biotechnology, Ranchi, India
‡CSIR-Institutes of Himalayan Bioresources and Technology, Palampur, India

4.1. INTRODUCTION

Lentil (*Lens culinaris* Medik.) is one of the most ancient legume crops; archeological evidence shows that they date back to 7500–6500 BC (Harlan, 1992). The scientific name of the lentil plant, *Lens culinaris*, was given by Medikus in the year 1787. The genus *Lens* belongs to family Fabacea (Leguminosae), subfamily Faboideae, tribe Vicieae, and it is grouped in the clade Hologalegina along with other cool-season legumes such as *Medicago*, *Pisum*, *Trifolium*, *Lathyrus* and *Vicia* (Gepts et al., 2005). *Lens culinaris* spp. *culinaris*, commonly called lentil, is the only cultivated species among the seven species of the genus (Van Oss et al., 1997). *Lens*, with a genome size of 4063 Mbps (Arumuganathan and Earle, 1991) and having *Lens culinaris* ssp. *orientalis* is its probable progenitor. Lentil is a self-pollinated diploid ($2n = 2\times = 14$) (Muehlbauer, 1991) crop, considered to have a Mediteranean center of origin (Cubero, 1981; Duke, 1981). Cultivated lentils are grouped into two subvarieties, namely, *macrosperma* and *microsperma*. The *macrosperma* has a seed size of more than 4 g of 100 seed weight and is grown mainly in the Mediterranean region and the New World, whereas *microsperma* is small-seeded, with a seed size less than 4 g for 100 seed weight and it is cultivated in the Indian subcontinent, Near East, and East Africa, respectively (Hawtin et al., 1980; Muehlbauer et al., 1985). The small-seeded lentil is insensitive to environmental stresses and photoperiods, whereas the large-seeded lentil is adapted to long days and sensitive to environmental stresses. Nevertheless, the latter possesses some useful genetic traits such as large seed size, upright and taller growth and resistance to various diseases. The small-seeded type is predominantly grown in India primarily due to its adaptation to shorter days. The *microsperma* lentil is one of the major pulse crops of the world, growing on an area of about 4.6 million hectares and producing

Lentils. https://doi.org/10.1016/B978-0-12-813522-8.00004-2

4.2 million tons of seeds with an average production of 1095 kg/ha (FAO, 2014). The crop is generally grown in rotation with cereals to break cereal disease cycles and to fix atmospheric nitrogen, reducing the demand for nitrogen fertilizers. Being a protein-rich crop, it provides more than twice the dietary protein of cereals. Its ability to thrive on marginal lands and drought-prone environments makes it a crop of choice of farmers in many developing countries. The distribution and cultivation area of lentils ranges from the Indian subcontinent to Northern Africa, Western Asia, Southern Europe, North and South America and Australia (Rubeena and Taylor, 2003). Thus, it is one of the major pulse crops of the world and requires systematic breeding and improvement programs to develop high-yielding varieties which should be resistant to various biotic and abiotic stresses. However, genetic variation and species relationships of the lentil germplasm are not well understood yet, making it impossible to utilize these for its improvement in the future. Genetic variations within populations of a crop species have been of major interest for plant breeders and geneticists for the introgression of desired genes from one species to other for improvement of a trait (Hayward and Breese, 1993). In the past, lentil breeding programs have relied mainly on local landraces of *Lens culinaris* ssp. *culinaris*. Thus, unlike other food legumes, lentil has lagged behind in terms of genetic improvement for yield and other major traits (Gupta and Sharma, 2006). Evaluation, screening and tagging of important genes of wild accessions can help in tapping unexplored variability and its probable use in lentil improvement programs. Therefore, in lentils, lots of wild resources are available which can be exploited not only to broaden the genetic base of cultivated lentils, but also to transfer desirable genes from other related wild species.

4.2. TAXONOMY

Lentils are known by as many as 30 common names in different parts of the world (Kay, 1979). The interrelationships among the cultivated and six related wild taxa have long been debated and most of the time, lentil breeders followed the distribution given by Ladizinsky (1979). Based on crossability of different *Lens* species, the genus *Lens* is grouped into main two species, namely, *Lens culinaris* and *Lens nigricans* (Ladizinsky et al., 1984a, b). *Lens culinaris* Medikus has three subspecies: *Lens culinaris* ssp. *culinaris, Lens culinaris* ssp. *orientalis*, and *Lens culinaris* ssp. *odemensis*, whereas *Lens nigricans* has two subspecies, namely, *Lens nigricans* spp. *nigricans* and *Lens nigricans* spp. *ervoides* (Ferguson et al., 2000). Initially, *Lens tomentosus* was considered the variant of *Lens orientalis* due to its morphological proximity, but in due course of time, Ladizinsky designated this new species as a *Lens tomentosus* Ladizinsky (Ladizinsky, 1979). The new order of distribution has been described as *Lens culinaris* having one cultivated type (*Lens culinaris* spp. *culinaris*) and three wild types (*Lens culinaris* ssp. *orientalis*, *Lens culinaris* ssp. *odemensis*, and *Lens culinaris* spp. *tomentosus*). With the identification of another wild species, *Lens lamottei*, the whole distribution of *Lens* species was reassessed by Van Oss et al. (1997). Sharma et al. (1995) differentiated the different lentil taxa representing wild and cultivated lentils using RAPD markers. *Lens culinaris* ssp. *orientalis* was the closest to the cultivated lentil, whereas *Lens ervoides* was the most divergent wild taxon, followed by *Lens nigricans*. The relationship between these two species was the same as between ssp. *orientalis* and the cultivated lentil. Sharma et al. (1996a, b) further confirmed the above findings based on AFLP markers, and Ahmad and McNeil (1996) compared a range of techniques for determining the interspecific relationships among lentil taxas. Van Oss et al. (1997) concluded that the genus *Lens* is comprised of seven taxas, namely, *Lens culinaris* ssp. *culinaris*, *Lens culinaris* ssp. *orientalis*, *Lens odemensis*, *Lens ervoides*, *Lens nigricans*, *Lens tomentosus*, and *Lens lamottei*. Ferguson et al. (2000) reclassified the phylogenetic relationship of the genus *Lens* based on crossability and morphological markers, which was further supported by the isozyme and RAPD data. Wong et al. (2015), on the basis of molecular characterization, redistributed the *Lens* species in detail. *Lens culinaris* is the only cultivated species, and two other species, *Lens orientalis* and *Lens tomentosus*, are closely related to cultivated lentils, and are grouped into the primary gene pool. *Lens odemensis* and *Lens lamottei* constitute the secondary gene pool, whereas *Lens ervoides* and *Lens nigricans* (which are considered very diverse and the species farthest from the cultivated variety) have been grouped into a tertiary gene pool (Fig. 4.1).

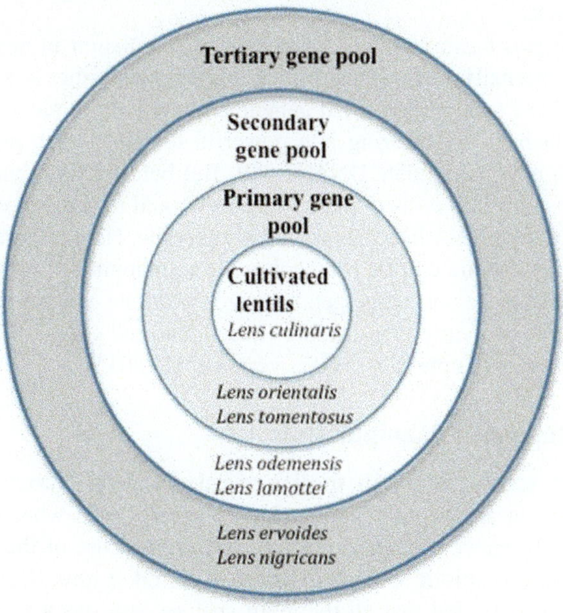

FIG. 4.1 Distribution of *Lens* genus (Wong et al., 2015).

Based on seed characteristics, Barulina (1930) subdivided the cultivated lentil (*Lens culinaris*) into two types, namely, *macrosperma* and *microsperma*. The *macrosperma* types were characterized by large seeds (6–9 mm diameter), yellow cotyledons and poor pigmentation on flowers and vegetative parts, whereas the *microsperma* types have small seeds (2–4 mm diameter), orange (red)/yellow cotyledons and pigmentation on flowers and vegetative parts. Although Williams et al. (1974) do not support the separation of two types of lentils based on seed size, but this concept is still largely accepted in lentil breeding programs.

4.3. MANIPULATION IN LENTILS

Genetic manipulation at the level of genotypes and genes has been practiced by humans for thousands of years by selecting desirable plant types and traits of interest. Genetic manipulation is a process for manipulating the genome of an organism in order to construct a more suitable ideotype. Initially, this was achieved by selective breeding, whereas modern genetic manipulation involves scientific procedures to add or delete DNA sequences in the target organism. There are numerous examples in history in which humans used this skill in selective breeding to select desirable plant types from naturally occurring genetic variations to produce new varieties, such as broccoli, cauliflower, cabbage and Brussels sprouts. Selective breeding is less effective now due to its inefficiency in selecting multiple desirable traits simultaneously. Modern genetic manipulation makes breeding plants with the desired traits more efficient, as it uses recombinant DNA technology to transfer an accumulation of desirable genes to a target plant.

There are two basic methods for genetic manipulation in any crop plant. One is conventionally known as selective breeding and other is a more recent and more refined technology called genetic engineering, or recombinant DNA technology. Genetic engineering is a powerful tool in human hands because scientists can now manipulate DNA at will. But before using this technology, scientists must develop tools to manipulate DNA and then develop methods to introduce modified DNA into the organism's genome. Hence, there is a need to develop efficient tissue culture protocol and fine mapping of all the desirable traits. The lentil has a very large genome size, so whole genome sequencing is still in progress, and various researchers have identified genes of interest using molecular markers (Gupta et al., 2012; Verma et al., 2015).

4.3.1 Cytogenetic Manipulation

Chromosome analysis of lentils species revealed that the chromosomal numbers are stable in all the cultivated and wild species, and whatever variation found in the chromosome morphology is due to exposure of these species to climatic, biotic and abiotic stresses over a long period of time. A critical review of the literature revealed that all the lentil species, sub-species, cultivars, and germplasms of *Lens* contain $2n = 2\times = 14$ chromosomes. There have been a number of cytological studies using various techniques in lentils, but there was no unanimity on chromosome morphology (metacentric, submetacentric, and acrocentric) karyotype formula, or total chromatin length (Bhattacharjee, 1951; Sharma and Mukhopadhyaya, 1963; Sinha and Acharia, 1972; Naithani and Sarbhoy, 1973; Gupta and Singh, 1981; Lavania and Lavania, 1983; Nandanwar and Narkhede, 1991). Ladizinsky (1993) used Feulgen staining and put all the controversy to rest by proposing that the *Lens* karyotype contains three types of chromosome; the first type is metacentric or submetacentric (centromere at or near the center). The second type is metacentric with a satellite, and the third type of chromosome is acrocentric (centromere is between the submetacentric and telomeric regions), and formulated with the 3 m/sm + 1 m with sat + 3 ac as a karyotypic formula. Balyan et al. (2002) also used modern techniques such as fluorescent in-situ hybridization (FISH) along with traditional Feulgen staining for chromosomal analysis of various *Lens* species, and observed that the position of secondary constriction in cultivated and some wild species, namely *Lens culinaris*, *Lens orientalis*, *Lens odemensis*, and *Lens ervoides*, was close to the centromere, that is, interstitial. However, in *Lens nigricans*, the position of secondary constriction was near the telomere of an acrocentric chromosome pair. Such cytological markers can be used to identify the hybrids as well introgression of desirable genes from *Lens nigricans*, but unfortunately, such methods have not been exploited to date in the lentil improvement programs. Application of fluorescent banding and FISH in the future may help to characterize apparently similar chromosomes at the constitutive heterochromatin level as an aid for future crop improvements.

Knowledge of cytological and molecular relationships between plant accessions is very useful in planning effective breeding strategies to transfer desirable genes or gene clusters from one species to another, thereby producing fruitful genomic reconstructions and disease free plants. Estimation of the genetic diversity of any crop species is a suitable precursor for improvement of the crop, because it generates baseline data to guide selection of parental lines and design of a breeding scheme. Genetic diversity refers to any variation in the nucleotides, genes, chromosomes, or genome of the organisms. Thus, each gene comprises a hereditary selection of DNA that occupies a specific place on the chromosomes and controls a particular characteristic of an organism (Welsh and McClelland, 1990).

4.3.2 Genetic Manipulations

Conventional manipulations mean the transfer of desirable gene or chromosome segments through natural hybridization. Most of the varieties developed worldwide arise only through intraspecific hybridization and pure line selection and have a narrow genetic base (Kumar et al., 2004). This makes them vulnerable to various biotic and abiotic stresses. Therefore, the introgression of useful genes from wild relatives has been suggested to overcome the problem of the narrow genetic base of lentils (Erskine et al., 1998; Rahman et al., 2009). Introgression not only beneficial for the transfer of useful genes from related wild species, but it also helps to broaden the genetic base of the cultivated lentils. It is a well-known fact that wild species are a rich reservoir of useful alien genes which are no longer available in the cultivated gene pool (Tanksley and McCouch, 1997). The genetic manipulation of lentils is primarily based on the exploitation of two gene pools of cultivated lentils (i.e., *macrosperma* and *microsperma*) by introgression of desirable genes from one another (Chahota et al., 1996; Chahota et al., 1997; Lal et al., 2000). Desirable trait production includes introgression of desirable genes from *macrosperma* with erect growth habits and biotic and abiotic stress-tolerance, and vice versa from *microsperma*, which has desirable traits like a higher number of branches, a higher number of seeds/plant, and a higher seed yield/plant. In lentil improvement programs, much has been reported about the creation of large amounts of variation following hybridization of the *microsperma* and *macrosperma* gene pools. Chahota et al. (2007) reported transgressive segregants for seed yield and other important agronomic traits for 77% of the crosses of *microsperma* x *macrosperma*. Though the main advantage of such hybridization programs is that two classes are easily crossable with no crossability barriers, these introgressions provide limited variation for improvement (Muench et al., 1991; Ferguson, 2000; Duran et al., 2004). In many crops, the wild relatives still possess important variation and are a source of desirable traits that no longer exist in their cultivated counterparts (Gupta and Sharma, 2007). To further broaden the genetic base of the two cultivated types of *micropsperma* and *macrosperma*, which are cultivated in different locations of the globe, wide hybridization is considered an alternative.

The wild progenitors of cultivated *Lens culinaris* spp. *orientalis* and *Lens culinaris* spp. *tomentosus* are in primary gene pools, so they can be crossed easily without any fertility problem (Fratini and Ruiz, 2006). The *Lens* gene pool consists of many wild relatives offering resistance to abiotic and biotic stresses (Ahmad et al., 1997; Hamdi et al., 1996). Continuous efforts have been made in the past in cultivated x wild hybridization, and there are some successful examples in which a few disease and insect pest-resistance genes have been successfully transferred to cultivated strains (Ladizinsky et al., 1988; Hjjar and Hodgpin, 2007; Fiala et al., 2009; Tullu et al., 2011). As far the crossability of wild *Lens* species with cultivated lentils is concerned, *Lens culinaris* spp. *orientalis* and *Lens culinaris* spp. *odomensis* are known to be crossable with cultivated lentils, so they share common gene pools (Ladizinsky, 1993). Gupta and Sharma (2007) attempted to cross *Lens culinaris* spp. *culinaris* with *Lens culinaris* spp. *orientalis* and *Lens culinaris* spp. *odomensis* and found normal meiosis and pollen fertility in F_1, F_2, and BC_1 crosses.

Many accessions of *Lens odemensis* and *Lens ervoides* showed drought tolerance, whereas, some accessions of *Lens culinaris* ssp. *orientalis* have been reported for cold tolerance and early maturity (Hamdi and Erskine, 1996; Hamdi et al., 1996; Gupta and Sharma, 2006). Bayaa and Erskine (1991) evaluated all

TABLE 4.1 Some Useful Traits Identified in Wild *Lens* Species (Bayaa and Erskine, 1991)

Trait	Species	Reference
Early flowering and maturity	*Lens culinaris* spp. *orientalis*	ICARDA (1991)
Pods/plant, Seeds/plants, straw yield and biomass	*Lens culinaris* spp. *orientalis, Lens nigricans,* and *Lens lamottei*	Ferguson and Robertson (1999)
100-seed weight	*Lens lamottei*	Ferguson and Robertson (1999)
Resistance to drought	*Lens odemensis, Lens ervoides*	Hamdi and Erskine (1996)
Winter hardiness	*Lens culinaris* spp. *orientalis*	Hamdi et al. (1996)
Resistance to vascular wilt	*Lens culinaris* spp. *orientalis, Lens ervoides*	Bayaa et al. (1995)
Ascochyta blight	All wild taxa	Bayaa et al. (1995)
Resistance to anthracnose	*Lens ervoides*	Fiala et al. (2009)

the ICARDA wild gemplasm for various diseases such as fusarium wilt caused by *Fusarium oxysporum* f.sp. *lentis* and ascochyta blight caused by *Aschochyta fabae* f.sp. *lentis* (Table 4.1). Combined resistance to ascochyta blight, fusarium blight, and anthracnose diseases have been reported in some of the *Lens ervoides* accessions (Bayaa et al., 1995; Tullu et al., 2006).

Pratap et al., 2014 evaluated 88 wild species of *Lens* and described the details of different morphological characters present in each wild accession. The summaries of results are presented in Table 4.2.

Fratini and Ruiz (2006), while studying the cross efficiencies of different species of *Lens* after ovule/embryo rescue, have reported that 90% of hybrids were recovered in the crosses of species present in secondary gene pools, namely, *Lens culinaris* spp. *culinaris* x *Lens culinaris* spp. *odemensis*, whereas, 3% of hybrid crosses were recovered when species are present in the tertiary gene pool, such as *Lens culinaris* spp. *culinaris* x *Lens nigricans* spp. *ervoides*. Gupta and Sharma (2006) have attempted crosses between cultivated *Lens culinaris* ssp. *culinaris* with two wild accessions of *Lens culinaris* spp. *orientalis* and *Lens culinaris* spp. *odomensis*; in crosses of *Lens culinaris* ssp. *culinaris* and *Lens culinaris* ssp. *orientalis*, normal seeds were produced, whereas varying degrees of chromosomal aberration and pollen fertility resulted while attempting crosses between *Lens culinaris* spp. *culinaris* x *Lens culinaris* spp.

TABLE 4.2 Morphophysiological Variation in Wild Accessions of Lentils

Trait	Mean	Range	Minimum	Maximum
Plant height (cm)	24.15	11.33–33.33	*Lens culinaris* spp. *orientalis*	*Lens ervoides*
Internode length	2.37	0.5–4.3	*Lens culinaris* spp. *orientalis*	*Lens ervoides*
Primary branches	3.71	1–7	*Lens ervoides*	*Lens culinaris* spp. *tomentosus*
Rachi's length	1.37	0.5–5.0	*Lens culinaris* spp. *orientalis*	*Lens nigricans*
Pods/cluster	1.35	1–3	*Lens culinaris* spp. *orientalis* *Lens ervoides* *Lens nigricans*	*Lens ervoides*
Leaf length (cm)	0.74	0.2–1.5	*Lens culinaris* spp. *orientalis*	*Lens culinaris* spp. *orientalis* *Lens ervoides* *Lens nigricans*
Leaf width (cm)	0.26	0.1–0.6	*Lens culinaris* spp. *orientalis* *Lens ervoides* *Lens nigricans* *Lens culinaris* spp. *tomentosus*	*Lens culinaris* spp. *orientalis*
Seeds/pod	1.66	1–3	*Lens ervoides*	*Lens nigricans*
100-seed weight	0.74	0.32–1.34	*Lens ervoides*	*Lens culinaris* spp. *tomentosus*

Source: Pratap, A., Kumar, J., Kumar, S., 2014. Evaluation of wild species of lentil for agro-morphological traits. Legum. Res. 37 (1), 11–18.

odemensis. Though the F_2 and BC_1 populations resulting from such crosses were evaluated for various agromorphological traits, they could not visualize the transfer of desirable traits to the cultivated lentils.

Lens nigricans contains two wild subspecies, *Lens nigricans* ssp. *nigricans* and *Lens nigricans* spp. *ervoides*. Taxonomically, these two spp. of *nigricans* are considered independent species and could not be crossed with *Lens culinaris* spp. *culinaris*, though many desirable traits are present in these two species, such as a higher number of pods/clusters, seeds/pod and 100-seed weight (Ferguson and Robertson, 1999). Under normal circumstances, these species cannot be crossed with *Lens culinaris* spp. *culinaris*, but using embryo rescue after 7–14 days of fertilization, F_1 and F_2 can be obtained (Cohen et al., 1984). It is further observed that when *Lens culinaris* spp. *culinaris* was taken as a female, different forms of chromosome rearrangements take place, the pollen stainability is 3%–5%, and no seed is produced. But when *Lens ervoides* was taken as a female, the pollen fertility increased to 52% and there was formation of some shriveled seeds. Out of the total seeds harvested, only 50% of seeds germinated and produced F_2 seeds. The segregation pattern was recorded to understand the inheritance pattern of some of these traits, such as epicotyl color and other seedling traits.

Lens tomentosus was initially considered a variant of *Lens culinaris* spp. *orientalis* on the basis of morphological proximity of the plants, but later on, Ladizinsky defined it a new species as the former was distinguishable by its hairy pod; it was further characterized by its modified kayrotypes and cpDNA restriction pattern (Van Oss et al., 1997). Although morphologically these *Lens tomentosus* are very close to the cultigen *Lens culinaris* spp. *culinaris*, interspecific hybrids can only be produced by employing embryo rescue technique. Suvorova (2014) successfully produced F_2-F_7 generation of *Lens culinaris* spp. *culinaris* x *Lens tomentosus* and was able to select lines from the segregation populations with good seed yield (1.51–1.78 t/ha) and higher protein content (27.4%). To date, there is no report of attempted hybridization involving *Lens Lamottei* with *Lens culinaris* spp. *culinaris*.

4.4. BREEDING STRATEGY

In breeding programs for self-fertilizing crops, a large number of crosses are being attempted every year in different parts of the world. To manage large numbers of recombinant lines in later generations, it would be highly desirable to start selection among crosses as early as possible. Developments in biometrical genetics have led to the suggestion that early generation trials may be used to predict the ranking of the crosses according to their likelihood of producing superior recombinant lines. Such cross prediction methods have been applied to a wide range of crop species such as barley, rice, potato, ryegrass, onion, sunflower, and oilseed rape. The F_2- or F_3-derived lines are far from being homozygous, and early generation selection relies on the assumption that the performance of a line at an early generation of selfing is predictive of its performance at homozygosity (Chahota et al., 2007). Empirical studies in different self-pollinated crops have indicated that early generation selection is sometimes effective and sometimes ineffective (Bernardo, 2003). In an autogamous crop like lentils, the identification of transgressive segregants in the form of inbred vigor for economically important traits such as seed yield is an important aspect of any practical breeding program. Considerable amounts of resources can be saved if there is objectivity about a particular cross likely to produce transgressive segregants in later generations. Frequency of transgressive segregants for a particular trait is based upon the distribution of positive genes in the parents, as described by Mather and Jinks (1982). This can be achieved by using precise estimates of various genetic components such as [d] (additive genic effect) and D (additive genetic variance), linkage among polygenes, and gene dispersion. If there is normal distribution of these components in the inbred lines, this probability integral will approximate to the two-tailed, normal probability integral corresponding with the value [d]/HD (Jinks and Perkins, 1972; Jinks and Pooni, 1976).

4.5. CONCLUSIONS

It can be concluded that, although there are many useful genes present in the wild species of *Lens*, due to the presence of strong crossability inhibitors, the introgression of these genes cannot be exploited to the fullest extent for the improvement of cultivated lentils. In some cases, the interspecific hybrids between *Lens culinaris* spp. *culinaris* with *Lens culinaris* ssp. *orientalis* and *Lens culinaris* spp. *odemensis* were produced, but in most cases, the endosperm was either lacking or abnormal. Hybrids of *Lens culinaris* spp. *culinaris* with *Lens nigricans* showed varying degrees of fertility due to problems in chromosome paring and chromosomal translocations (Goshen et al., 1982; Ladizinsky et al., 1984a, b). Similarly, crosses of *Lens culinaris* spp. *culinaris* with *Lens tomentosus* were not successful due to chromosome pairing, resulting in abortion of the embryo (Ladizinsky, 1979).

REFERENCES

Ahmad, M., McNeil, D.L., 1996. Comparison of crossability, RAPD, SDS-PAGE and morphological markers for revealing genetic relationships within and among *Lens* species. Theor. Appl. Genet. 93, 788–793.

Ahmad, M., McNeil, D.L., Sedcole, J.R., 1997. Phylogenetic relationships in *Lens* species and their interspecific hybrids as measured by morphological characters. Euphytica 94, 101–111.

Arumuganathan, K., Earle, E.D., 1991. Nuclear DNA content of some important plant species. Plant Mol. Biol. Report. 9 (3), 208–218.

Balyan, H.S., Houben, A., Ahne, R., 2002. Karyotype analysis and physical mapping of 18S-5.8 S-25S and 5S ribosomal RNA loci in species of genus *Lens* Miller (Fabaceae). Caryologia 55 (2), 121–128.

Barulina, O.H., 1930. Lentil of USSR and of other countries. Bull. Appl. Bot. Genet. Plant Breed. 40, 1–319.

Bayaa, B., Erskine, W., 1991. Legume Program—Annual Report. ICARDA, Aleppo, Syria.

Bayaa, B., Erskine, W., Hamdi, A., 1995. Evaluation of a wild lentil collection for resistace to vascular wilt. Genet. Resour. Crop. Evol. 42, 231–235.

Bernardo, R., 2003. On the effectiveness of early generation selection in self-pollinated crops. Crop Sci. 43, 1558–1560.

Bhattacharjee, S.K., 1951. Cytogenetics of *Lens esculenta* Moench microsperma. Sci. Cult. 16, 426–427.

Chahota, R.K., Gupta, V.P., Sharma, S.K., 1996. Combining ability and genetic architecture of *macrosperma* x *microsperma* derived crosses. J. Hill Res. 9 (2), 341–346.

Chahota, R.K., Sharma, S.K., Gupta, V.P., 1997. Heterotic and inbreeding effects in lentil crosses involving *macrosperma* and *microsperma* (*Lens culinaris* Medik.). Crop. Improv. 24 (1), 74–77.

Chahota, R.K., Kishore, N., Dhiman, K.C., Sharma, T.R., Sharma, S.K., 2007. Predicting transgressivesegregants in early generation using single seed descent method-derived micro-macrospermagenepool of lentil (*Lens culinaris* Medikus). Euphytica 156 (3), 305–310.

Cohen, D., Ladizinsky, G., Ziv, M., Muehlbauer, F.J., 1984. Rescue of interspecific *Lens* hybrids by means of embryo culture. Plant Cell Tissue Organ Cult. 3 (4), 343–347.

Cubero, J.I., 1981. Origin, domestication and evolution. In: Webb, C., Hawtin, G.C. (Eds.), Lentils. Commonwealth Agricultural Bureau, Slough, UK, pp. 15–38.

Duke, J.A., 1981. Handbook of Legumes of World Economic Importance. Plenum Press, New York52–57.

Duran, Y.R., Fratini, P., Garcia, Perez de la Vega, 2004. An intersubspecific genetic map of *Lens*. Theor. Appl. Genet. 108, 1265–1273.

Erskine, W., Chandra, S., Chaudhry, M., Malik, I.A., Sarker, A., Sharma, B., Tufail, M., Tyagi, M.C., 1998. A bottleneck in lentil: widening its genetic base in South Asia. Euphytica 101 (2), 207–211.

FAO, 2014. Faostat, Fao Statistical Database. Retrieved from: http://www.fao.org.

Ferguson, M., 2000. Conserved resources, priorities for collection and future prospects. In: Knight, R. (Ed.), Lining Research and Marketing Opportunities for Pulses in 21st Century, Current Plant Science and Biotechnology in Agriculture. vol. 34. pp. 613–620.

Ferguson, M.E., Maxted, N., Slageren, M.V., Robertson, L.D., 2000. A re-assessment of the taxonomy of *Lens* Mill. (Leguminosae, Papilionoideae, Vicieae). Bot. J. Linn. Soc. 133, 41–59.

Ferguson, M.E., Robertson, L.D., 1999. Morphological and phenological variation in the wild relatives of lentil. Genet. Resour. Crop Evol. 46, 3–12.

Fiala, J.V., Tullu, A., Banniza, S., Séguin-Swartz, G., Vandenberg, A., 2009. Interspecies transfer of resistance to anthracnose in lentil (Medic.). Crop Sci. 49 (3), 825–830.

Fratini, R., Ruiz, M.L., 2006. Interspecific hybridization in the genus *Lens* applying in vitro embryo rescue. Euphytica 150 (1), 271–280.

Gepts, P., Beavis, W.D., Brumones, E.C., Shoemaker, R.C., Stalker, H.J., Weeden, N.F., Young, N.D., 2005. Legumes as a nodule plant family. Genomics for food and feed report of the cross legume advances through genomics conference. Plant Physiol. 137, 1228–1235.

Goshen, D., Ladizinsky, G., Muehlbauer, F.J., 1982. Restoration of meiotic regularity and fertility among derivatives of *Lens culinaris* x *Lens nigricans* hybrids. Euphytica 31 (3), 795–799.

Gupta, D., Sharma, S.K., 2006. Evaluation of wild *Lens* taxa for agro-morphological traits, fungal diseases and moisture stress in north western Indian Hills. Genet. Resour. Crop. Evol. 53, 1233–1241.

Gupta, D., Sharma, S.K., 2007. Widening the gene pool of cultivated lentils through introgression of alien chromatin from wild Lens subspecies. Plant Breed. 126 (1), 58–61.

Gupta, P.K., Singh, J., 1981. Standard karyotype in lentil (*Lens culinaris* Med.) var. PL-639. Lens 8, 23.

Gupta, M., Verma, B., Kumar, N., Chahota, R.K., Rathour, R., Sharma, S.K., Bhatia, S., Sharma, T.R., 2012. Construction of intersubspecific molecular genetic map of lentil based on ISSR, RAPD and SSR markers. J. Genet. 91 (3), 101–110.

Hamdi, A., Erskine, W., 1996. Reaction of wild species of the genus *Lens* to drought. Euphytica 91, 173–179.

Hamdi, A., Kusmenoglu, Erskine, W., 1996. Sources of winter hardiness in wild lentil. Genet. Resour. Crop Evol. 43, 63–67.

Harlan, J.R., 1992. Crops and Man, second ed. American Society of Agronomy, Madison, Wisconsin, USA284.

Hawtin, G.C., Singh, K.B., Saxena, M.C., 1980. Some recent developments in the understanding and improvement of *Cicer* and *Lens*. In: Summerfield, R.J., Bunting, A.H. (Eds.), Advances In Legume Science. pp. 613–623.

Hayward, M.D., Breese, E.L., 1993. Population structure and variability. In: Hayward, M.D., Bosemark, N.O., Romayosa, I. (Eds.), Plant Breeding: Principles and Prospects. Chapman and Hall, London, pp. 7–29.

Hjjar, R., Hodgpin, T., 2007. The use of wild relatives in crop improvement: a survey of development over the last 20 years. Euphytica 156, 1–13.

Jinks, J.L., Perkins, J.M., 1972. Predicting the range of inbred lines. Heredity 28, 399–403.

Jinks, J.L., Pooni, H.S., 1976. Predicting the protperites of redombinatn inbred lines derived by single seed descent. Heresdity 36, 253–266.

Kay, D.E., 1979. Good Legumes, Tropical Production Institute, Crop and Production Digest, No. 3, London.

Kumar, S., Gupta, S., Chandra, S., Singh, B.B., 2004. How wide is the genetic base of pulse crop? Pulses in new perspective. In: Ali, M., Singh, B.B., Kumar, S., Dhar, V. (Eds.), Indian Society of Pulses Research and Development. Indian Institute of Pulses Research, Kanpur, India, pp. 211–222.

Ladizinsky, G., 1979. The origin of lentil and its wild gene pool. Euphytica 28, 179–187.

Ladizinsky, G., 1993. Wild lentils. Crit. Rev. Plant Sci. 12, 169–184.

Ladizinsky, G., Braun, D., Goshen, D., Muehlbauer, F.J., 1984a. The biological species of the genus *Lens* L. Bot. Gaz. 145, 253–261.

Ladizinsky, G., Cohen, Muehlbauer, F.J., 1984b. Hybraidization in the geuns *Lens* by means of embryo culture. Theor. Appl. Genet. 70, 97–101.

Ladizinsky, G., Pickersgill, B., Yamamoto, K., 1988. Exploitation of wild relatives of the food legumes. In: Summerfield, R.J. (Ed.), World Crops, Cool Season Food Legumes. Kluwer, Dordrecht, pp. 967–987.

Lal, C., Sharma, S.K., Chahota, R.K., 2000. Combining ability and heterosis for growth parameters in *microsperm* x *macrosperma* lentil. Crop. Res. 19 (1), 115–121.

Lavania, U.C., Lavania, S., 1983. Karyotype studies in Indian pulses. Genet. Agrarian 37 (3–4), 279–307.

Mather, K., Jinks, J.L., 1982. Biometrical Geneitcs, the Study of Continuous Variation, third ed. S. Champman and Hall, London, New York396.

Muehlbauer, F.J., 1991. Use of introduced germplasm in cool season food legume cultivar development. In: Shands, H.L., Wiesner, L.E. (Eds.), Use of Plant Introductions in Cultivar Development (Part 2). Crop Science Society America, pp. 49–73. Social publication no. 20.

Muehlbauer, F.J., Cubero, J.I., Summerfield, R.J., 1985. Lentil. In: Summerfield, R.J., Roberts, E.H. (Eds.), Grain Legume Crops. Collins, London, pp. 266–311.

Muench, D.G., Slinkard, A.E., Scoles, G.J., 1991. Determination of genetic variation and taxonomy in lentil (*Lens* Miller) species by chloroplast DNA polymorphisms. Euphytica 56, 213–218.

Naithani, S.P., Sarbhoy, R.K., 1973. Cytological studies in *Lens esculenta* Moench. Cytologia 38 (2), 195–203.

Nandanwar, R.S., Narkhede, M.M., 1991. Intraspecific variation in karyotype of lentil. J. Maharashtra Agric. Univ. 16, 24–27.

Pratap, A., Kumar, J., Kumar, S., 2014. Evaluation of wild species of lentil for agro-morphological traits. Legum. Res. 37 (1), 11–18.

Rahman, M.M., Sarker, A., Kumar, S., Ali, A., Yadav, N.K., Rahman, L., 2009. Breeding for short season environments. The lentil: botany, production and uses. In: Erskine, W., Muehlbauer, F., Sarker, A., Sharma, B. (Eds.), CAB Int. pp. 121–136. Wallingford, UK.

Rubeena, F.R., Taylor, P.W.J., 2003. Construction of an intraspecific linkage map of lentil (*Lens culinaris* ssp. *culinaris*). Theor. Appl. Genet. 107, 910–916.

Sharma, A.K., Mukhopadhyaya, S., 1963. Karyotype constancy in different strains of *Lens esculentum* Moench as worked out through recent techniques. Indian Agric. 7, 103–111.

Sharma, S.K., Dawson, I.K., Waugh, R., 1995. Relationships among cultivated and wild lentils revealed by RAPD analysis. Theor. Appl. Genet. 91, 647–654.

Sharma, S.K., Knox, M.R., Ellis, T.H.N., 1996a. AFLP analysis of diversity and phylogeny of *Lens* and its comparison with RAPD analysis. Theor. Appl. Genet. 93, 751–758.

Sharma, D.R., Kaur, R., Kumar, K., 1996b. Embryo rescue in plants—a review. Euphytica 89 (3), 325–337.

Sinha, S.S., Acharia, S.S., 1972. Karyotype analysis in some varieties of *Lens culinaris*. Cytologia 37, 673–683.

Suvorova, G., 2014. Hybridization of cultivated lentil (*Lens culinaris* Medik.) and wild species *Lens tomentosus* Ladizinsky. Czech J. Genet. Plant Breed. 50 (2), 130–134.

Tanksley, S.D., McCouch, R., 1997. Seed banks and molecular maps: unlocking genetic potential from the wild. Sciences 277, 1063–1066.

Tullu, A., Buchwaldt, L., Lulsdorf, M., Banniza, S., Barlow, B., Slinkard, A.E., Sarkar, A., Taran, T.D., Warkentin, T.D., Vanderberg, A., 2006. Souces of resistance to anthracnose (*Colletotrichum truncatum*) in wild *Lens* species. Genet. Resour. Crop. Evol. 53, 111–119.

Tullu, A., Banniza, S., Bett, K., Vandenberg, A., 2011. A walk on the wild side: exploiting wild species for improving cultivated lentil. Grain Legume 56, 13–14.

Van Oss, H., Aron, Y., Ladizinsky, G., 1997. Chloroplast DNA variation and evolution in the genus *Lens* Mill. Theor. Appl. Genet. 94, 452–457.

Verma, P., Goyal, R., Chahota, R.K., Sharma, T.R., Abdin, M.Z., Bhatia, S., 2015. Construction of a genetic linkage map and identification of QTLs for seed weight and seed size traits in lentil (*Lens culinaris* Medik.). PLoS ONE. https://doi.org/10.1371/journal.pone.013966.

Welsh, J., McClelland, M., 1990. Fingerprinting genomes using PCR with arbitrary primers. Nucleic Acids Res. 18, 7213–7218.

Williams, J.T., Sanchez, A.M.C., Jackson, M.T., 1974. Studies on lentils and their variation: 1. The taxonomy of the species. SABRAO J. Breed. Genet. 6 (2), 133–145.

Wong, M.M., Gujaria-Verma, N., Ramsay, L., Yuan, H.Y., Caron, C., Diapari, M., Vandenberg, A., Bett, K.E., 2015. Classification and characterization of species within the genus *Lens* using genotyping-by-sequencing (GBS). PLoS ONE 10 (3), e0122025.

FURTHER READING

Thavarajah, P., Thavarajah, D., 2011. The potential of lentil (*Lens culinaris* L.) as a whole food for increased selenium, iron, and zinc intake: preliminary results from a 3 year study. Euphytica 180 (1), 123–128.

Chapter 5

Embryo Rescue and Cytogenetic Manipulation

Dorin Gupta*, Rebecca Ford†, Prabhakaran Sambasivam†, Sajitha Biju*

**School of Food and Agriculture, Faculty of Veterinary and Agricultural Sciences, University of Melbourne, Dookie, VIC, Australia, †Environmental Futures Research Institute, School of Natural Sciences, Griffith University, Nathan, QLD, Australia*

5.1. INTRODUCTION

Lentil (*Lens culinaris* ssp. *culinaris* Medikus, $2n = 2X = 14$) belongs to tribe *Vicieae*, family Fabaceae (Leguminoseae), and is a diploid self-pollinating legume. The crop provides both environmental and ecological benefits within an agricultural system, including the contribution of nitrogen fixation and breaking cereal crop pathogen and insect pest cycles. The lentil grain contains more than twice the amount of dietary protein (25%), essential amino acids, minerals, and carbohydrates, and is an integral part of the diet of poor people, for whom accessibility of animal protein is limited (Srivastava and Vasishtha, 2012). Among the world's oldest crops, lentils remain an integral part of agricultural production systems. *L. culinaris* originated in the Near East around the Fertile Crescent and was later domesticated in southern Turkey with further spread to the Nile, Greece, Europe, and Asia. *L. culinaris* ssp. *orientalis* (Boiss.) Ponert is the presumed wild progenitor of the cultivated lentil and is found from western Greece to eastern Uzbekistan and the northern Crimean Peninsula to southern Jordan (Renfrew, 1969; Ladizinsky, 1979; Cubero, 1984). The occurrence of wild lentils around the Mediterranean basin suggests its popularity during the early agricultural settlement followed by its domestication, local adaptation, and further spread around the world, leading to emergence of geographically diverse landraces (Erskine, 1997). However, over time, South Asia became the world's largest lentil growing area (approximately 48%), with little variation for the lentils of a specific ecotype (*pilocea*) growing in this region, which remained a major limitation for lentil breeding progress (Erskine et al., 1998).

Lentils are a cool season food legume cultivated in the Indian subcontinent, Canada, Australia, southern Europe, eastern and northern Africa, and the dry areas of the Middle East. Although ancient, the crop has not been exploited to its full potential as reflected by yield plateaus, and the breakdown of disease and

Lentils. https://doi.org/10.1016/B978-0-12-813522-8.00005-4

insect resistance. Huge selection pressure for developing new varieties and narrow genetic diversity within the cultigen (Ford et al., 1997) are mainly responsible for the crop's vulnerability to co-evolved pathogens (Podder et al., 2013). In a preferred cereal-based cropping system, lentils have been a crop of marginal lands, which might have contributed to loss of genes for higher productivity and hence further narrowing of the genetic base (Bejiga and Degago, 2000). To improve crop performance, variation should exist in the cultivated gene pool from which desirable genetic selections may be made. When this is not the case, in order to widen this gene pool, distant cultivated and wild relatives are relatively untapped sources of variations for high yields and resistance to stresses which can either be introduced into new regions, hybridized with existing varieties, or utilized in mutation breeding (Erskine et al., 1998). Although wild relative genomes may be used to improve and widen the genetic base of the cultivated gene pool, genetic and cytogenetic manipulations may be required. For this, considerations of genome size, composition, and compatibility/incompatibility leading to fertility barriers must be made.

5.2. TAXONOMY OF THE *LENS* GENUS

Lentil has a fairly large genome of about 4 Gbp with seven chromosome pairs (Arumuganathan and Earle, 1991). Cultivated lentils are broadly categorized into two classes—small-seeded (microsperma) and large-seeded (macrosperma) based on morpho-physiological traits. Wong et al. (2015) recently classified seven lentil species into four gene pools, in which *L. culinaris*/*L. orientalis*/*L. tomentosus* form primary, *L. lamottei*/*L. odomensis* form secondary, *L. ervoides* forms tertiary and *L. nigricans* forms quaternary gene pools (Fig. 5.1). Among the seven species, the genus *Lens* encompasses only one cultivated lentil species (*L. culinaris*), and the remaining six are wild species. Furthermore, the wild species *L. orientalis* is presumed to be the wild progenitor of cultivated lentils

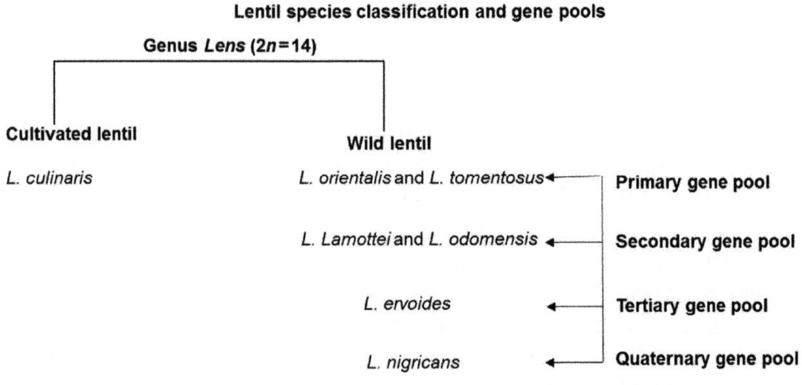

FIG. 5.1 Genus *Lens* and seven species categorized into four gene pools based on studies by Wong et al. (2015).

(Zohary, 1972). Wild species from primary and secondary gene pools are easily crossable with cultivated lentil, unlike with the wild species from the remaining two gene pools (Gupta and Sharma, 2007; Singh et al., 2013).

Traditional and molecular cytogenetic techniques have been employed to identify introgressed chromosomal segments among broad crosses (Goshen et al., 1982; Ladizinsky and Abbo, 1993). These approaches have also been used to aid in fertility restoration for intra/interspecific hybrids (Ahmad et al., 2005; Gupta and Sharma, 2007; Singh et al., 2013). Cytogenetic manipulations include biochemical and molecular biology approaches such as in chromosome banding, fluorescence in situ hybridization (FISH), and genome in situ hybridization (GISH) (Gall and Pardue, 1969; Rayburn and Gill, 1985; Pinkel et al., 1986; Le et al., 1989). These cytomolecular techniques have greatly facilitated the precise study of chromosomes (heterochromatin, centromeres, and telomeres) and their putative functions. In addition, molecular cytogenetics have played a vital role in genome analysis and sequencing of important crops to utilize distant relative genetic resources and combat the risk of genetic erosion by improving the chances of wide hybridization to introgress genes of interest into cultigens such as rice (Cheng et al., 2002), wheat (An et al., 2015; Zhang et al., 2017), brassica (Fredua-Agyeman et al., 2014), and banana (D'Hont et al., 2000).

Cytogenetic analysis of lentil chromosomes enabled an understanding of lentil genome evolution, particularly, species chromosome number and structure. Recently, a clear differentiation and evolution of chromosomes within *Lens* speciation was reported (Galasso, 2003; Fernández et al., 2005; Gaffarzadeh-Namazi et al., 2007). The future application of these techniques will be to facilitate the transfer of durable gene(s) of interest for biotic and abiotic stresses, and/or agronomical traits of high yield with reduced linkage drag from distant/wild species to provide new sources of variation. Furthermore, embryo-ovule rescue has been used to improve introgression of wild species chromatin into cultivated ones from noncrossable gene pools. The wide hybrids have been studied cytogenetically to ascertain their role in lentil breeding programs. Advanced cyto-molecular/genomic techniques have provided a clearer picture of introgressed chromosomes and/or chromosome arms, and aided in the association of physical and genetic maps and mapping of chromosome-specific markers. The creation of aneuploids is another method to introduce novel genetic variation in order to create new genetic combinations for desirable traits. These can be generated through mutagenesis in higher plants (Nagar, 2013) including lentils (Kumar and Gupta, 1998; Gupta et al., 1999), chickpeas (Pundir et al., 1983), and green grams (Sharma, 1990).

5.3. CYTOGENETICS OF THE *LENS* GENUS

Traditional cytogenetic techniques were employed in the 20th century to understand the lentil genome. Bhattacharjee (1953) studied lentil meiotic chromosomes in detail using oxyquinoline. However, by that time, it was already known that there were six species within genus *Lens* with a basic chromosome number (*n*), which varied from five to eight. He reported the diploid chromosome number of lentil $2n = 14$, which was in agreement with the earlier reports of Heitz (1926), Bleier (1928), and Miranda (1931), and for karyotype analysis with Bhattacharjee (1951). He concluded that the number of nucleoli corresponded with the number of secondary constrictions, which were a maximum of four in the somatic chromosomes. However, the meiotic study also revealed considerable irregularities in the behavior of the chromosomes and irregular disjunctions leading to pollen grain development with chromosome numbers greater or less than seven. Based on these findings, he suggested that the plant under study was not a true diploid (Bhattacharjee, 1953).

For almost 30 subsequent years, classical cytogenetic manipulations were the key drivers to further clarifying the number and status of wild *Lens* species, the wild progenitor of cultivated lentils, and the chromosomal structures and number of the various *Lens* species. Ladizinsky (1979) greatly expanded the knowledge in these areas using morphological and cytogenetic studies. He suggested that *Lens* comprised five annual species, four being wild (*L. montbretii*, *L. ervoides*, *L. nigricans*, and *L. orientalis*) with only *L. culinaris* being the cultivated species. He further suggested that *L. nigricans*, *L. orientalis*, and *L. culinaris* had the same chromosome number ($2n = 14$). He further confirmed the findings of Barulina (1930) and Williams et al. (1974), suggesting that the origin of cultivated lentil (*L. culinaris*) was *L. orientalis*. He also confirmed the similarity of the *L. culinaris* karyotype to fourteen *L. culinaris* lines studied by Sharma (1962). However, Williams et al. (1974) suggested a different karyotype for *L. culinaris*, and Ladizinsky subsequently reported that *L. orientalis* genotypes were cytogenetically uniform but differed from *L. culinaris* by a single chromosome interchange (Ladizinsky, 1979). Meanwhile, *L. nigricans* and *L. culinaris* cytogenetically differed by three chromosome interchanges (the presence of a large number of univalents, rod bivalents, trivalents, quadrivalents, and pentavalents) (Ladizinsky, 1979). However, their hybrids were fertile, indicating a monophyletic origin (Ladizinsky, 1979). Subsequently, Ladizinsky et al. (1984) proposed two species of *Lens* (*L. culinaris* and *L. nigricans*). Among these two species, *L. culinaris* included cultivated lentil (*L. culinaris* ssp. *culinaris*) and *L. culinaris* ssp. *orientalis* (wild progenitor of cultivated lentil) and *L. culinaris* ssp. *odomensis*.

In another study, the F_1 progeny of a *L. culinaris* x *L. nigricans* cross showed irregular chromosome pairing due to the differences of three interchanges among the parents (Goshen et al., 1982). Only 19% of the F_2 plants were meiotically stable and possessed chromosomal arrangements similar to that of cultivated lentil (Goshen et al., 1982). Slinkard (1985) studied karyotypes of all lentil species and suggested that seven pairs of chromosomes have median or near-median centromeres with a secondary constriction near the centromere of one or two pairs of chromosomes. Based on these findings and chromosome paring of F_1, he confirmed Ladizinsky's findings that *L. culinaris* and *L. nigricans* form two crossable groups with subspecies of *L. culinaris* spp. *culinaris*, *L. culinaris* spp. *orientalis*, *L. culinaris* spp. *odomensis*, *L. nigricans* spp. *nigricans*, and *L. nigricans* spp. *ervoides*. Slinkard's (1985) classical cytogenetic and crossing results suggested the genus *Lens* should be considered one large gene pool from a breeding viewpoint, but as two separate gene pools for evolutionary consideration.

5.3.1 Cytogenetic Manipulations and Ploidy Status

The 1980s saw some remarkable work in classical cytogenetics and exploration of the wild species. Among various cytogenetic manipulations, polyploidy revolutionized monocots and dicots, and was considered an important crop improvement tool. The application of colchicine was a common method to artificially induce polyploidy, through chromosome doubling, also known as colchiploidy. Malaviya and Shukla (1983) undertook cytogenetic analysis of lentil cultivars to understand the presence and extent of ploidy level. They revealed three triploid and two tetraploid plants among 100 plants of variety WYR134 and one triploid plant in ILL467. Cytogenetic studies of these polyploids exhibited irregular meiotic events such as abnormal chromosomal synapsis, non-oriented chromosomes at metaphase 1, the presence of laggards, unequal chromosome numbers at the poles, and a high percentage of pollen sterility. They also supported the theory of a stabilized diploid level in *Lens.* One year later, Malaviya and Shukla (1984) reported cytological features of six colchiploid plants. They concluded that the developed colchiploids were not stable and observed tetraploid and mixoploids (pollen mother cells with 29 and 27 chromosomes) as well as irregular chromosomal behavior during different meiotic stages, which could further explain why natural polyploids do not exist in lentils. Understanding and utilizing constancy of nuclear DNA content of a species through cytogenetic manipulation helps to understand the evolutionary relationships (Rees, 1984). Considerable nuclear DNA content variation among interspecific individuals has been reported by various researchers for different crops including several legumes (Schweizer and Davis, 1972; Furuta, 1975; Laurie and Bennett, 1985). Ramesh (1994) looked for variation in chromatin content, nuclear DNA amount, and distribution of DNA among different chromosomes and also performed somatic karyotyping of ten cultivated lentil genotypes. They concluded that intervarietal variation was significant for chromatin and nuclear DNA content. Metacentric chromosomes were the highlight of the symmetric lentil karyotypes, with only one chromosome pair detected with secondary constrictions near to the centromere in only two of ten varieties. The 2C DNA content was positively correlated with total chromosome volume and nuclear area, and DNA density was negatively correlated with nuclear area. Ramesh (1994) consequently suggested that the changes in DNA content with genotypic evolution have affected all chromosomes at different magnitudes. In summary, using traditional cytogenetic approaches, many contradictory reports arose on the exact lentil karyotype (Bhattacharjee, 1953; Ladizinsky, 1979; Sharma, 1962; Williams et al., 1974; Ladizinsky et al., 1984; Goshen et al., 1982; Slinkard, 1985; Malaviya and Shukla, 1983, 1984). More advanced cytogenetic technologies were subsequently employed to more accurately identify the *Lens* chromosomes and establish robust karyotypes as follows.

5.3.2 Chromosome Banding Techniques

Kannan and Zilfalil (2009) reviewed and summarized various cytogenetic techniques. Among the banding techniques, Q-banding was the first (Caspersson et al., 1970) which involved the use of fluorochrome (quinacrine mustard or quinacrine dihydrochloride) to stain chromosomes followed by their examination under fluorescence microscopy. The G-banding technique (Rowley, 1973) became popular as this had a better resolution than Q-banding, where chromosomes were stained with Giemsa solution. Giemsa C-banding technique was also developed (Pardue and Gall, 1970), which darkly stains the constitutive heterochromatin regions of chromosomes, and helps to identify polymorphic regions of chromosomes and multiple centromeres of a chromosome. As the name implies, another technique, nucleolar organizing region (NOR)-banding or N-banding (Matsui and Sasaki, 1973) stains chromosomal NORs. This can localize centromeric heterochromatin and identify the NORs of SAT chromosomes in eukaryotic genomes that give rise to the nucleolus (and thus ribosomes), and identifying their exact location on a chromosome is of great significance for precise genome analysis. Mehra et al. (1986) used N-banding for karyotype analysis of lentils and to locate satellite (SAT) chromosomes. Near metacentric chromosomes were reported to be large, whereas, the small chromosomes were submetacentric. Further, they inferred that the SAT chromosome had comparatively small, short arms between the primary and secondary constrictions. These findings suggested the SAT chromosome as an acrocentric or sub-telocentric chromosome. Mehra et al. (1986) explained that the existence of the large satellite was responsible for the appearance of the SAT chromosome as a sub/medium-metacentric chromosome detected in earlier studies (Sinha and Acharia, 1972; Ladizinsky, 1979). The power of this new cytogenetic technique clarified Sinha and Acharia's (1972) findings of not being able to locate the lentil SAT chromosomes using simple cytogenetic techniques and conventional staining; hence, the two constrictions might have appeared as one complete section. However, the karyotypic analysis using the N-banding technique (Mehra et al., 1986) was in agreement with that reported by Sinha and Acharia (1972), that lentils possessed large metacentric chromosomes, and that the SAT and small chromosomes were submetacentric. Meanwhile, the Mehra et al. (1986) finding differed from that of Ladizinsky (1979) who suggested that lentils possess a SAT chromosome as medium metacentric and a small chromosome as acrocentric.

Gaffarzadeh-Namazi et al. (2007) used C-banding to overcome gaps in the karyotypic knowledge of lentil chromosomes and identified individual lentil chromosomes which had four pairs of metacentric and three pairs of sub-metacentric chromosomes. They also assessed C-banding polymorphism among lentil lines and reported each chromosome to have a unique C-banding pattern. Also, they reported that the long arm of Chromosome 4 has a secondary constriction near the centromeric region, which was in agreement with earlier studies (Slinkard, 1985; Raziuddin et al., 1990; Ahmad et al., 1992). Gaffarzadeh-Namazi et al. (2007) also agreed with Kumar and Gupta (1997) and Galasso et al. (2001) in identifying the SAT chromosome as the fourth largest chromosome among the seven. This differed from Ahmad et al. (1992), who reported that the SAT chromosome was the second longest.

5.3.3 Cytogenetic and Mutation Studies

Mutation to increase genetic variation can be induced artificially to develop new varieties in less time than spontaneous mutation (Micke et al., 1990). Various chromosomal aberrations arise due to mutagen treatment including chromosomal interchanges/reciprocal translocations (exchange of non-homologous chromosome segments). Ionizing radiation has been used in various plant species to produce such interchanges (Gill et al., 1980). These interchanges can be employed as cytogenetic markers which play an important role in creating new variations and basic material for cytological studies (Burnham, 1932). From a plant cytogeneticist's viewpoint, interchanges with known chromosome involvement are valuable, particularly when a tester set can be made from many interchanges (Gupta and Gupta, 1991; Gupta, 1995) for use in gene mapping (Mahama et al., 1999). Burnham (1962) and Gupta (1995) suggested analysis of somatic chromosomes, analysis of pachytene configurations, semi-sterility of genetic markers, meiotic analysis of the F_1s developed through crossing of known trisomics and interchanges, and the crosses between unknown interchanges as methods to identify these potential interchanges. Understanding pairing configurations during meiosis provides insight into chromosome structure rearrangement, which is a significant contribution to understanding the plant's evolution (King et al., 1993). Also, there are some successful examples of important trait transfers in crops like wheat and barley using reciprocal translocations (Driscoll and Sears, 1965).

Kumar and Gupta (1998) were the first to report the isolation of a complex translocation monosomic lentil plant from the F_2 generation of a cross of two translocation homozygotes with different chromosomes. They found interchange heterozygotes carrying two trisomics ($2n = 2x + 1 = 15$) from the progeny of the monosomic. Simple microscopic meiotic studies of these plants revealed the presence of multivalents, which reflected the presence of multiple interchanges, possibly in five of the seven non-homologous chromosomes of the monosomic. This may help to explain why trisomic plants do not exist, since one of the studied plants had abnormal morphological traits (compact branching, narrow leaves, and dwarf nature) due to an extra chromosome. Gupta et al. (1999) described the significance of a standard interchange tester set for a plant species to identify individual chromosomes involved in interchanges, or aneuploids by studying crosses with the testers. Among different methods to identify individual chromosomes, they used meiotic analysis of the diallel cross (interchange homozygotes) hybrids, and chromosome morphology alterations to identify individual chromosomes. An interchange tester set can be utilized for chromosome mapping, and more specifically, for the identification of particular chromosomes, which might be involved in an interchange. To create this, Gupta et al. (1999) irradiated the seed of cultivar PL-639 with different doses of gamma rays and produced many chromosome interchanges. From studies of meiotic configurations at metaphase I and somatic chromosomes, they identified 14 interchange homozygotes through altered chromosome morphology, including seven interchange homozygotes assembled in the tester set. Nine of these 14 interchange homozygotes had easily identifiable interchanged chromosomes due to changed total lengths and the ratios of long and short arms of each chromosome. In a recent study, Goyal and Verma (2015) irradiated lentil seeds with different doses of gamma rays, and with different concentrations of ethyl methane sulphonate (EMS). They selected two mutant plants, one of which was a translocation heterozygote (gamma rays treated), and the other, an inversion heterozygote (EMS treated). At various meiotic stages, the plants showed different types of chromosomal configurations in most of the pollen mother cells such as the ring/chain formation of four and six chromosomes of the translocation heterozygote. Meanwhile, the inversion heterozygote bridged and contained fragments which were attributed to the crossover number and position in the inverted segment. In general, though, the inversion heterozygote had poor pollen fertility compared to the translocation heterozygote (Goyal and Verma, 2015).

In summary, there are few reports in which cytogenetics has played a role to understand changes in the lentil genome through tracking of induced mutations. This approach has not been used anywhere near the extent it has in other major crops. This leaves a large scope for future studies to understand these variations more precisely and to assess their potential for use in lentil mutation breeding programs.

5.3.4 Molecular Cytogenetics

Molecular cytogenetic approaches involving FISH and GISH of highly repetitive DNA sequences, restriction site analyses and nucleotide variations have proven to be a valuable tool in the more precise physical visualization of chromosomes. These are also used in the identification of structure, function, organization, and localization of chromosomes, genes and DNA sequences. These methods involve hybridization of DNA probes to denatured metaphase spreads, followed by direct or immunological staining. The probes used in FISH can be region-specific, including ribosomal DNA (rDNA), arm-specific, and chromosome or genome-specific. By employing this method, one can identify with accuracy the chromosomes involved in the organization of the nucleolus and the exact location of genes. They can also be used to detect numerical changes in chromosomes and to compare the genomes of interspecific or intergeneric hybrids (Lombello and Pinto-Maglio, 2004). FISH and its modifications, such as multiple-target fluorescence in situ hybridization (multicolor FISH) (Mukai, 1996), extended DNA fiber-FISH (Ersfeld, 2004), bacterial artificial chromosome (BAC)-FISH (Hanson et al., 1995), and super-stretched pachytene-FISH (Koo and Jiang, 2009), have been employed to reveal the minute details of chromosome structure and subsequently permit sophisticated analyses of chromosomal behavior (Sharma et al., 2016).

GISH is a technique used to identify alien chromosomes as well as chromatin and chromosomal rearrangements formed due to mosaic chromosomes (Ramzan et al., 2017). GISH distinguishes genomic relationships within polyploids. Also analysis of genome composition, visualization, and chromosome discrimination from different genomes in allopolyploids are carried out by GISH (Singh, 2003). GISH techniques follow the same procedure as FISH, but genomic and blocking DNA are utilized in GISH. These two techniques have been employed in crop improvement programs, molecular systematics, and in the conservation and utilization of plant genetic resources (Hizume et al., 2002; Chaudhary et al., 2011). They have been employed to unravel individual chromosomes in depth specifically in legumes like chickpea (Khattak et al., 2007), soybean (Findley et al., 2010, 2017), medicago (Pires et al., 2012), garden peas (Greilhuber and Ebert, 1994), faba beans (Caperta et al., 2008), sweet pea (Seijo and Fernández, 2003) and peanut (Robledo et al., 2009). However, these molecular chromosomal visualization techniques have not been so extensively employed in lentil. Most of the studies in lentils have focused on physical and genetic comparative mapping (Bhattacharjee, 1951; Sharma, 1962; Naithani and Sarbhoy, 1973; Gupta and Singh, 1981; Nandanwar and Narkhede, 1991; Tullu et al., 2011; Jha et al., 2015).

Abbo et al. (1994) used FISH to localize the nucleolus organizer region (NOR) in somatic chromosomes of *L. culinaris* and detected the hybridization sites of an rDNA probe coding for the 18S, 5.8S, and 26S genes. One conserved pair of hybridization sites of a ribosomal probe were detected among *L. culinaris* and *L. orientalis*, which corresponded to the secondary constriction site. Galasso et al. (2001) subsequently used FISH to characterize the chromosomal location of two repeated DNA sequences (pLc30: 466 bp long with 64% AT residues, and pLc7: 408 bp long with 61% AT residues). They reported two families with sequence differences, where the pLc30 family had four internal repeats and the pLc7 family had short direct sub-repeats. When they looked into the presence of these families in various *Lens* species and other genera, pLc30 was found only in lentil species; however, pLc7 was present in other genera. The FISH results further showed that pLc30 sequence hybridized on six of seven chromosomes, whereas, pLc7 hybridized with only one chromosome pair. A more robust identification of the seven chromosome pairs was proposed based on nine different hybridization sites of the pLc30 family. This led to assembly of a physical map based on FISH results with pLc7, 5S rRNA, 18S–5.8S–25S rRNA genes, and a telomeric sequence. They demonstrated that in situ hybridization with these two probes (pLc30 and pLc7) allowed the discrimination of all seven chromosome pairs and enabled the construction of a *L. culinaris* standard FISH karyotype (Galasso et al., 2001).

Meanwhile, Kumar and Ramesh (2001) provided evidence for two pairs of chromosomes carrying 18S–5.8S–25S rRNA loci by FISH using the interchange heterozygote (where a satellited chromosome is involved in the interchange) and decondensed prophase chromosomes. Balyan et al. (2002) also used Feulgen staining as well as fluorescent in situ hybridization for chromosomal analysis of five different species of *Lens* and observed that the position of the secondary constriction in cultivated and some wild species, namely *L. culinaris*, *L. orientalis*, *L. odomensis*, and *L. ervoides*, was close to the centromere, that is, interstitial. However, in *L. nigricans*, the position of secondary constriction was near the telomere of an acrocentric chromosome pair. Thus, the five species had similar karyotypes with minor modifications, which may be due to the chromosomal interchanges of chromosomal fragments of similar size during the evolution of the *Lens* genus.

Balyan et al. (2002) summarized the work of various researchers from the 1950s to the early 1990s (Bhattacharjee, 1951; Sharma, 1962; Sinha and Acharia, 1972; Naithani and Sarbhoy, 1973; Keshwani, 1973; Gupta and Singh, 1981; Sharma and Gupta, 1982; Lavania and Lavania, 1983; Nandanwar and Narkhede, 1991), who reported zero to three SAT chromosomes in cultivated lentils. However, in their own study, they concluded that the 18S–5.8S–25S rRNA genes and 18S–5.8S–25S rDNA (the intergenic spacer regions) are tandem repeats specifically present at the NORs and possibly localized at other chromosomal sites without connection with NORs. This was suggested by earlier studies (Mukai et al., 1991; Leitch and Heslop-Harrison, 1992; Jiang and Gill, 1994; Pedersen and Linde-Laursen, 1994). Balyan et al. (2002) suggested that since the 5S rRNA genes are present at one or more loci per chromosome set and occur in tandem repeats, these together with 18S–5.8S–25S may be detectable by FISH. Balyan et al. (2002) subsequently successfully mapped 18S–5.8S–25S rRNA at a single locus on a single chromosome pair and 5S rRNA at two loci with varying sizes of two different repeat units on two different chromosome pairs for each of the five species. They further suggested the use of the rRNA loci as landmarks for chromosome identification, which were found at different positions on chromosomes in different species. Their studies concluded that the karyotypes of all five species which was considered to be the same, were in fact different on the idiograms and the rRNA loci physical map.

Galasso (2003) conducted multiple-target fluorescence in situ hybridization studies (multicolor FISH) for the simultaneous detection of multiple targets to study the repeated DNA sequences in *Lens* species. For this, two highly repetitive sequences (pLc30 and pLc7) isolated from the cultivated lentil and the multigene families for the 18S–5.8S–25S (pTa71) and 5S rRNA (pTa794) from wheat were used simultaneously as probes. Each species showed a typical FISH karyotype, and few differences were observed among accessions belonging to the same species, except for the accessions of *L. odomensis*. The FISH karyotype most similar to the cultivated lentil was that of *L. orientalis*, whereas *L. nigricans* and *L. tomentosus* showed the most divergent FISH patterns compared with all taxa based on number and location of pLc30 and 18S–5.8S–25S rDNA sites. Sonnante et al. (2003) then isolated the internal transcribed spacer (ITS) region of the nuclear ribosomal DNA of *Lens* sp. They reported sequence divergence of 14 base substitutions between *L. nigricans* and *L. lamottei*, but close phylogeny among cultivated lentil to *L. orientalis* (wild progenitor) at this region. They proposed *L. lamottei* and *L. tomentosus* as separate species based on unusual autapomorphies.

Other than the FISH technique, restriction endonuclease digestion of DNA and the squash technique have also been employed to understand evolutionary relationships in *Lens*. Mayer and Soltis (1994) studied chloroplast DNA (cpDNA) variation using 399 restriction-sites to determine relationships among 114 cultivated and 11 wild accessions. They also used this approach to ascertain the wild progenitor of cultivated lentil and to construct a cpDNA phylogeny. Most of the accessions had the same cpDNA except three from the cultivated genome. Among the wild accessions, they identified restriction-site mutations and length mutations and confirmed *L. orientalis* as the wild progenitor. Based on limited cpDNA diversity, they were also able to conclude the possible loss of cytoplasmic variability during domestication. Based on ten autapomorphic restriction-site mutations, they also determined that *L. nigricans* was the most divergent taxon of the genus.

Recently, Pal et al. (2016) used the squash technique to study the somatic chromosome and used sodium dodecyl sulfate polyacrylamide gel electrophoresis (SDS-PAGE) for seed protein profiling of five lentil varieties. Karyotype analysis highlighted small differences for each variety producing individual karyotypes, and protein profiling revealed an expected low diversity.

5.4. WIDE HYBRIDIZATION, EMBRYO-OVULE RESCUE AND HAPLOID PRODUCTION

5.4.1 Embryo-Ovule Rescue

Plenty of evidence points to wild lentils as a useful source of untapped genetic variations for important agronomical traits, disease resistances, and abiotic stress tolerances. Several attempts have been made to introgress desirable traits from wild lentils to the cultivated gene pool with the eventual aim of selecting promising recombinants from backcrossing to advanced cultivated accessions (Gupta and Sharma, 2006, 2007; Fiala et al., 2009; Gupta et al., 2012; Tullu et al., 2013; Singh et al., 2013). Indeed, introgressed populations have shown transgressive segregants which surpass their parents in those traits, indicating the creation of new recombinant progeny that may be utilized in breeding programs. Among cultivated and wild lentil species, crosses are easy to make within the primary, and to some extent, between the primary and secondary gene pools. (Gupta and Sharma, 2007; Singh et al., 2013). Although some genotypic variation for the success of crossability with cultivated lentils exists even within these closer gene pools (Gupta and Sharma, 2007; Singh et al., 2013, 2017), the search for significant variation for selective purposes may need to go much wider and hence into the tertiary and quaternary gene pools. These more distantly related gene pools contain more desirable chromatin for stress resistances, which is lacking or easily broken in cultivated varieties (Gupta and Sharma, 2007; Singh et al., 2013; Tullu et al., 2013; Dadu et al., 2017).

Interspecific hybrids of cultivated x *L. orientalis* (primary gene pool) and *L. odomensis* (secondary gene pool) have been successfully developed through conventional hybridization techniques (Vandenberg and Slinkard, 1989; Fratini et al., 2004; Gupta and Sharma, 2007; Singh et al., 2013). However, the hybrid fertility may be affected by altered chromosomal arrangements (Ladizinsky et al., 1984; Gupta and Sharma, 2007). A far reduced success rate is achieved in obtaining fertile interspecific hybrids from crossing between cultivated lentil and wild species from the tertiary and quaternary gene pools such as *L. ervoides* x *L. nigricans*. For example, Goshen et al. (1982) studied interspecific hybrids of cultivated x *L. nigricans* crosses and found irregularities in meiotic analysis of the hybrids arising from the difference between parental chromosomes for three translocations. Conversely, cytological analysis of 14 populations of cultivated x *L. orientalis* crosses by Ladizinsky and Abbo (1993) revealed similar chromosome compositions as cultivated lentils. Hamdi and Erskine (1994), Gupta (2003), and Kumari et al. (unpublished) selected promising progenies from a bulk segregating population of cultivated x *L. orientalis* or *L. odomensis* crosses that exhibited high variability and high yields. However, the same authors had little success when crossing cultivated lentil with species from the tertiary and quaternary gene pools, likely due to pre- and postfertilization barriers (Gupta and Sharma, 2005). Prefertilization barriers exist mostly due to asynchronous flowering which may be rectified with staggered sowing. On the other hand, postfertilization barriers are mostly due to embryo failure (Gupta and Sharma, 2005), which may be resolved through embryo-ovule rescue techniques. Indeed, conventional hand-pollinated crossing and in vitro rescued embryos remains the only feasible approach for successful interspecific hybrid recovery in many cases (Bridgen, 1994; Sharma et al., 1996).

An embryo-ovule rescue technique is used to overcome interspecific fertilization barriers due to hybrid embryo abortion (Sharma et al., 1996). The first evidence of hybrid embryo abortion of such interspecific crosses in *Lens* was evident from the findings of Ladizinsky et al. (1985). They reported shrivelled non-viable seeds from wide cross-derived pods, which ceased development 10–16 days after pollination. This is one of the major impediments to uninterrupted gene flow from these wild species into cultivated lentils. To solve this, the embryo-ovule rescue technique was first used to create hybrids between cultivated lentils *L. nigricans* and *L. ervoides* (Ladizinsky et al., 1984). Abbo and Ladizinsky (1991) similarly observed hybrid embryo abortion 14 days after pollination in cultivated lentila x *L. ervoides* and upon examination of the aborted embryos, reported the presence of small cotyledons, shoot and root primordia, and endosperm remnants. They subsequently optimized the embryo rescue techniques to obtain vegetatively normal hybrids.

Abbo and Ladizinsky (1994) shed new light in understanding the cause of embryo abortion and suggested that chromosomal aberrations might not be the main reason for F_1 embryo abortion of *L. orientalis* x *L. culinaris* and *L. ervoides* x *L. culinaris* cross combinations. However, they did suggest a strong effect of dominant gene action on embryo abortion, and both dominant and additive gene action was inferred from segregating populations. Ahmad et al. (1995) successfully crossed cultivated lentil with *L. orientalis*, *L. odomensis*, *L. ervoides* and *L. nigricans*. This was achieved using the plant growth regulator gibberellic acid (GA_3) with success rates from 50% to 100%, but with much lower rates of healthy hybrid seeds. Further in vitro culturing of hybrid embryos that did not grow was unsuccessful with death after 3 weeks. Ye et al. (2002) then developed a seed culture based protocol for efficient in vitro propagation of hybrid seed and obtained multiple shoots with the addition of cytokinins such as benzyl adenine (BA), thidiazuron (TDZ), and kinetin. They reported that Murashige and Skoog (MS) media was more optimal for shoot induction than Gamborge (B5) basal media. Among the three cytokinins used, BA and TDZ produced more shoots than kinetin and a higher calcium concentration was beneficial to avoid shoot-tip necrosis. They also successfully induced rooting using MS medium supplemented with Naphthaleneacetic (1-NAA) for more than 50% of the shoots obtained. Later, in order to rescue wide hybrid embryos, a two-step in vitro method was followed by Gupta and Sharma (2005), in which ovules were rescued to obtain viable hybrids of cultivated lentil crossed with *L. ervoides* and *L. nigricans* (Fig. 5.2). For this, they used the embryo culture method described by Cohen et al. (1984) with some modifications to the culture media. Ovules of about two-week-old selfed and hybrid crosses were transferred to standard Murashige and Skoog medium (MS stock solutions) + 100 g sucrose + 0.5 mg/L GA3 + 0.5 mg/L BAP + 0.2 mg/L IAA + 6 g/L agar. Later, these were moved for shoot-regeneration to MS media containing BAP (0.5–1.0 mg/L), kinetin (0.5–1.0 mg/L), IAA (0.2 mg/L), sucrose (30 g/L) and agar (6.0–8.0 g/L). These were kept at $25 \pm 1°C$ in the dark for 3 days and then placed under 12 h/12 h daylight/dark. After 7 days, the embryos were excised from the ovules and placed into test tubes containing MS media with eight different types and concentrations of growth-hormones, amino acid sources, and agar. This was done to optimize the MS medium suitable for embryo growth and further differentiation into shoots and roots. Although they observed single and multiple shoot differentiation on the hybrid embryos, the embryos failed to differentiate roots (Fig. 5.3).

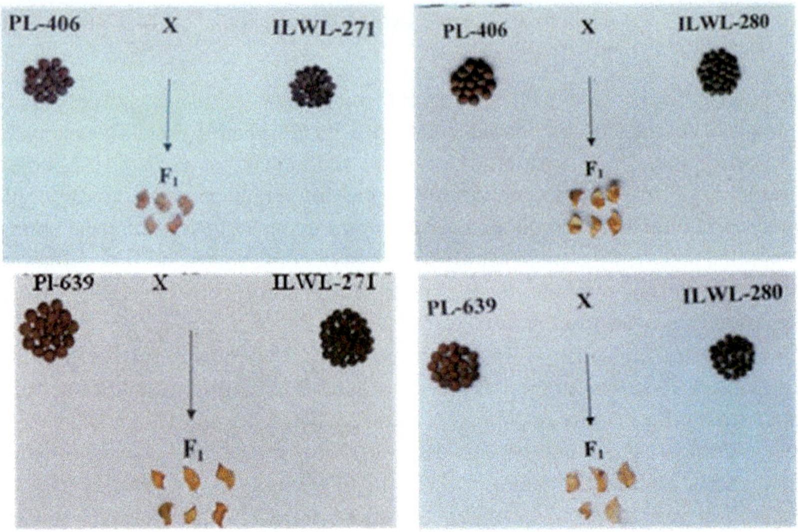

FIG. 5.2 Shrivelled pods and immature embryos due to embryo abortion of crosses among cultivated and wild lentil genotypes.

FIG. 5.3 (A) Green callus, (B) brown callus, (C) shoot primordia, and (D) single shoot regeneration of interspecific hybrid ovules rescued after 14 days of pollination

In 2006, Fratini and Ruiz reported obtaining the first successful interspecific hybrids from a cultivated x *L. odomensis* cross following Cohen et al.'s (1984) embryo rescue protocol. They also proposed a new protocol where ovules were first cultured after 18 days of pollination. The medium was comprised of MS Salts, sucrose (1%), indole acetic acid (1 μM), and Kinetin (0.8 μM). Once ovules had been established in vitro for 2 weeks, the embryos were excised from the ovules and maintained on the same medium. They successfully obtained interspecific plantlets of cultivated x *L. odomensis* (6), as well as cultivated x *L. nigricans* (2) and cultivated x *L. ervoides* (1) and concluded a rescue efficiency range of 50% to 100% based on hybrid recovery per number of ovules cultured. Later, Fratini and Ruiz (2011) published an embryo-ovule rescue method and compared this with Cohen's method and suggested that the difference in the efficacy of the two methods was due to the media and carbohydrate content used. Cohen et al. (1984) used 10% and 3% of sucrose in the media and Fratini and Ruiz (2006) only used 1% sucrose. One of the most successful attempts to transfer genes of interest from wild to cultivated lentils using the embryo culture technique was reported by Fiala et al. (2009), who used this method to transfer resistance for two races of anthracnose from *L. ervoides* to cultivated lentils. Individual F_2 plants were then raised to obtain $F_{7:8}$ recombinant inbred lines through single-seed descent (SSD).

Transgenic plant technology, which offers the opportunity to transfer genes of interest from wider sources, is also reliant on successful in vitro culturing of transgenic explants. A regeneration protocol which is efficient and reliable is an important prerequisite for genetic transformation success. Sarker et al. (2003) developed an agrobacterium-mediated transformation system for two lentil microsperma varieties to test the transformation ability of different explants (cotyledonary node, decapitated embryo, immature embryo, and epicotyl). They reported GUS positive regions with the highest percentage for epicotyl explants followed by decapitated embryo. However, they found decapitated embryo to be most useful for multiple shoot formation on MS medium + 0.5 mg/L BAP + 0.5 mg/L Kn + 0.1 mg/L GA3 + 5.5 mg/L tyrosine and with stable expression of the GUS gene in transformed shoots. Recently, the use of a small portion of the embryo axis with the cotyledon as the explant has been suggested as an efficient lentil regeneration protocol (Tavallaie et al., 2011). Bagheri et al. (2012) studied different hormone treatments and explants to determine the best reproducible protocol for in vitro regeneration of an Iranian cultivar. Among 13 different hormone treatments and 4 explants for callus induction and regeneration, modified MS medium containing 1 mg/L α-NAA and 1 mg/L Zeatin (medium E) and decapitated embryos attached to 1/4 of the cotyledon produced the highest fresh and dry weight callus with highest shooting and rooting responses (75%). Recently, Suvorova (2014) successfully used an ovule rescue technique to obtain hybrids from cultivated lentil x *L. tomentosus* (ILWL 90 and ILWL120) crosses. Three hybrids were recovered from 296 rescued ovules. Hybrids and F_2s developed only through the ovule rescue technique from a cross of cultivated and ILWL 90 were either sterile or partly fertile. However, hybrids of cultivated and ILWL 120 crosses were fertile whether obtained through ovule rescue or without ovule rescue, indicating a strong genotypic effect. They were successful in generating breeding recombinant lines of high productivity through selection of plants in F_2–F_7 generations of this cross. However, concern was raised about the taxonomic status of ILWL 120 as a member of *L. tomentosus* due to the ease of crossability.

To overcome the rooting issue, Yuan et al. (2011) tested the feasibility of in vivo intergeneric grafting techniques using faba bean as a rootstock for lentil scions. When an accession of each of six wild *Lens* species was used as the scion in grafts to faba bean breeding line FB50-9 rootstock, successful grafts were obtained for all species with survival to seed maturity rates of between 70.7% and 87.7% except for the *L. orientalis* accession (55.3% survival). Days to flowering and seed parameters were the same between control and grafted plants, indicating this technique might be useful to considerably speed up efficient interspecific lentil hybrid production with unique wild genetic resources. Khentry et al. (2014) reported successful in vitro propagation of six lentil genotypes (Digger, Indianhead, Nipper, Northfield, ILL 7537, and ILL 6002). In vitro regenerated nodes were transferred to MS medium + 4 mg/L α-naphthalene acetic acid which resulted into about 80% of the regenerated plantlets be normal plants under in vivo conditions. Their finding was useful to report the efficient establishment of roots of normal morphology. More recently, Saha et al. (2015) used an in vitro embryo rescue approach followed by in vivo grafting of the rescued embryo shoots onto faba bean root stocks for the successful production of wide *Lens* hybrids. Accordingly, they rescued embryos from 14-day-old immature hybrid seeds and cultured them on a medium containing auxin 4-chloroindole-3 acetic acid (4-Cl-IAA) and zeatin combinations that led to proliferation of shoots, and they did not observe any significant difference in shoot elongation for 4-Cl-IAA verses IAA treatment. Subsequent in vivo grafting of hybrid shoots onto faba bean root stocks resulted in large numbers of F_2 seed from crosses between cultivated lentil with *L. tomentosus*, *L. lamottei*, or *L. odomensis*.

In summary, regarding shoot and root regeneration from in vitro cultured lentil explants, rooting presents the greatest challenge. Many researchers have reported none or low rooting frequency as an impediment in the regeneration of in vitro cultures (Ye et al., 2002; Sarker et al., 2003; Khawar et al., 2004; Gupta and Sharma, 2005). Mohamed et al. (1992) suggested that cytokinin used in shooting culture media may be associated with no or poor rooting initiation, and this required further exploration. Meanwhile, the use of related genera such as the faba bean looks to be a promising alternative.

Embryo culture also has potential to shorten the breeding cycle as was demonstrated by Gatti et al. (2016) who developed an in vitro-in vivo system using embryo culture along with an SSD method to shorten the breeding cycle in lentil variety development. They collected embryos of microsperma and macrosperma seed types sat 15, 18, 21, and 24 DAP and cultured these on MS medium with five different concentrations of BAP ranging from 0 to 0.25 mg/L. Genotypes, DAP, and their interaction had significant effects on the percentage of germination and shoot formation and indicated that microsperma genotypes are better respondents with an optimized 18 DAP embryo rescue timing for highest shoot production and germination percentage. Furthermore, BAP did not negatively affect embryo-to-plant development, with the ability to produce four generations per year of normal and fertile plants for rapid recombinant inbred lines.

5.4.2 Haploid Breeding

For the fast recovery of fixed lines of desired hybrids, conventional methods used in breeding programs need evaluation and selection of traits of interest in several segregating generations. To reduce the time required, the doubled-haploid (DH) approach has been applied, estimated to shorten varietal production time by five years because homozygous lines may be produced in a single generation (Pratap et al., 2005; Croser et al., 2006; Pratap et al., 2010). This technique utilizes the inherent potential of the cells be able to develop in a haploid plant from either the male or female gametes (n) only. Generally, androgenesis leads to production of haploids from microspores directly or by culturing anthers with microspores (Croser et al., 2006).

The first attempts to assess the potential of haploids were as early as 1964 and 1966, when Guha and Maheshwari (1964, 1966) reported the production of plants using Datura anther culture. Another powerful technique for obtaining barley haploids was demonstrated by Kasha and Kao (1970) using the chromosome elimination technique following distant hybridization. Every crop for which a haploid has been successfully produced has its own protocol, specifically for in-vitro and prepreparation requirements. Croser et al. (2006) have reviewed and summarized attempts done so far to standardize various steps and protocols for different legumes. They reported that haploid production using wide hybridization methods had been used successfully in cereal and potato crops, whereas microspore culture offers potential for producing haploids for most species. They also pointed out the issues that impede efforts in legumes, including their genetic disposition to low numbers of microspores per anther. Therefore, although this technology has tremendous potential to increase the efficacy of breeding programs, there are few successful attempts to produce doubled-haploids in grain legumes (Zhao and Sharp, 1998; Kaur and Bhalla, 1998; Mix and Wang, 1988). Genotypic differences, regeneration conditions, and poor or no in-vitro rooting are some of many factors impacting DH production in lentils (Croser et al., 2006) and wide hybrids (Gupta and Sharma, 2005). Keller and Ferrie (2002) attempted to develop lentil haploidy technology by testing various lentil genotypes with different environments, media, and culture approaches. They also explored many pretreatment conditions, anther culture, and appropriate microspore stages to instigate embryogenesis.

Also, researchers at the Crop Development Centre in Saskatoon, Canada, and the Australian Centre for Legumes in Mediterranean Agriculture have attempted to produce lentil doubled-haploids using wide hybridization and tissue culture techniques such as anther culture and isolated microspore culture (https://grdc.com.au/research/reports/report?id=267). Croser and Lulsdorf (2004) reported that one to five days of pretreatment at 4°C triggered an induction response for microspore culture from two Canadian lentil genotypes (CDC Crimson and CDC Robin). However, this led to early embryo development with no plant regeneration (Croser and Lulsdorf, 2004; Croser et al., 2006). Most recently, the GRDC of Australia reported the groundbreaking success of one putative lentil haploid from crosses performed between lentil and field pea and from many wide (interspecific and intergeneric) hybridization attempts (https://grdc.com.au/research/reports/report?id=267). In summary, anther and microspore cultures are suggested as better options than interspecific hybridization for the successful production of lentil double-haploids. However, this will require more species- and genotype-specific research and funding to support scientific progress toward development of a widely implementable protocol.

5.5. CONCLUSIONS

To date, improvement for higher yields and resistance to major diseases and pests in lentils has been retarded by a lack of financial investment in R&D. Although most breeding programs around the world recognize that their genetic base is narrow and that wild relative species offer a wealth of genetic material, there has been a lack of investment in the advanced technologies required to enable the movement of this material into the elite primary cultivated gene pool.

Distant hybridization (aided with biotechnological tools like embryo-ovule rescue), together with modern cytogenetic manipulations, has a lot to offer (including the ability to precisely karyotype) to the development of intra- and interspecific genetic populations, genetic maps, association maps, quantitative trait loci, marker-assisted selection implementation in the lentil breeding program, and eventual fast-tracking of the deployment of genes responsible for desirable traits. Also, the application of GISH and FISH in the future may help to characterize apparently similar chromosomes at the constitutive heterochromatin level to aid in future lentil breeding. These technologies are urgently required to fully realize the significance of legumes such as lentils in our sustainable production systems, to guarantee the food security of millions around the world.

REFERENCES

Abbo, S., Ladizinsky, G., 1991. Anatomical aspects of hybrid embryo abortion in the genus *Lens* L. Bot. Gaz. 152 (3), 316–320.

Abbo, S., Ladizinsky, G., 1994. Genetical aspects of hybrid embryo abortion in the genus *Lens* L. Heredity 72 (2), 193–200.

Abbo, S., Miller, T.E., Reader, S.M., Dunford, R.P., King, I.P., 1994. Detection of ribosomal DNA sites in lentil and chickpea by fluorescent in situ hybridization. Genome 37 (4), 713–716.

Ahmad, H., Khan, I.M., Alam, M., Khawaja, H.I.T., 1992. Karyotypic studies in *Lens culinaris*. Pakphyton 4, 181–189.

Ahmad, M.A.G.F., Fautrier, A.G., McNeil, D.L., Burritt, D.J., Hill, G.D., 1995. Attempts to overcome post fertilization barrier in interspecific crosses of the genus *Lens*. Plant Breed. 114 (6), 558–560.

Ahmad, F., Gaur, P.M., Croser, J., 2005. Chickpea (*Cicer arietinum* L.). In: Singh, R., Jauhar, P. (Eds.), Genetic Resources, Chromosome Engineering and Crop Improvement-Grain Legumes. vol. 1. CRC Press, Boca Raton, FL, pp. 185–214.

An, D., Zheng, Q., Luo, Q., Ma, P., Zhang, H., Li, L., Han, F., Xu, H., Xu, Y., Zhang, X., Zhou, Y., 2015. Molecular cytogenetic identification of a new wheat-rye 6R chromosome disomic addition line with powdery mildew resistance. PLoS ONE 10 (8), e0134534.

Arumuganathan, K., Earle, E.D., 1991. Nuclear DNA content of some important plant species. Plant Mol. Biol. Report. 9 (3), 208–218.

Bagheri, A., Ghasemi Omraan, V.O., Hatefi, S., 2012. Indirect *in vitro* regeneration of lentil (*Lens culinaris* Medik.). J. Plant Mol. Breed. 1 (1), 43–50.

Balyan, H.S., Houben, A., Ahne, R., 2002. Karyotype analysis and physical mapping of 18S-5.8 S-25S and 5S ribosomal RNA loci in species of genus *Lens* Miller (Fabaceae). Caryologia 55 (2), 121–128.

Barulina, O.H., 1930. Lentil of USSR and of other countries. Bull. Appl. Bot. Genet. Plant Breed. 40, 1–319.

Bejiga, G., Degago, Y., 2000. Region 4: Sub-Sahara Africa. In: Knight, R. (Ed.), Linking Research and Marketing Opportunities for Pulses in the 21st Century. Springer, Netherlands, pp. 99–105.

Bhattacharjee, S.K., 1951. Karyotype analysis of *Lens esculenta*. Sci. Cult. 16, 426–427.

Bhattacharjee, S.K., 1953. Cytogenetics of *Lens esculenta* Moench: (With Plate XIV and 11 figures). Caryologia 5 (2), 159–166.

Bleier, H., 1928. Karyologische Untersuchungen an Linsen-Wicken-Bastarden. Genetica 11 (1), 111–118.

Bridgen, M.P., 1994. A review of plant embryo culture. Hortscience 29 (11), 1243–1246.

Burnham, C.R., 1932. An interchange in maize giving low sterility and chain configurations. Proc. Natl. Acad. Sci. 18 (6), 434–440.

Burnham, C.R., 1962. Discussions in Cytogenetics. Burgess, Minneapolis.

Caperta, A.D., Rosa, M., Delgado, M., Karimi, R., Demidov, D., Viegas, W., Houben, A., 2008. Distribution patterns of phosphorylated Thr 3 and Thr 32 of histone H3 in plant mitosis and meiosis. Cytogen. Genome Res. 122 (1), 73–79.

Caspersson, T., Zech, L., Johansson, C., 1970. Differential binding of alkylating fluorochromes in human chromosomes. Exp. Cell Res. 60 (3), 315–319.

Chaudhary, H.K., Sood, V.K., Tayeng, T., Kaila, V., Sood, A., 2011. Molecular cytogenetics in physical mapping of genomes and alien introgressions. In: Biology and Breeding of Food Legumes. CABI, Oxfordshire, pp. 131–146.

Cheng, Z., Buell, C.R., Wing, R.A., Jiang, J., 2002. Resolution of fluorescence in-situ hybridization mapping on rice mitotic prometaphase chromosomes, meiotic pachytene chromosomes and extended DNA fibers. Chromosom. Res. 10 (5), 379–387.

Cohen, D., Ladizinsky, G., Ziv, M., Muehlbauer, F.J., 1984. Rescue of interspecific *Lens* hybrids by means of embryo culture. Plant Cell Tissue Organ Cult. 3 (4), 343–347.

Croser, J.S., Lulsdorf, M.M., 2004. In: Progress towards haploid division in chickpea (*Cicer arietinum* L.), field pea (*Pisum sativum* L.) and lentil (*Lens culinaris* Medik.) using isolated microspore culture. European Grain Legume Conference. AEP, Paris, Dijon, p. 189.

Croser, J.S., Lülsdorf, M.M., Davies, P.A., Clarke, H.J., Bayliss, K.L., Mallikarjuna, N., Siddique, K.H.M., 2006. Toward doubled haploid production in the Fabaceae: progress, constraints, and opportunities. Crit. Rev. Plant Sci. 25 (2), 139–157.

Cubero, J.I., 1984. Taxonomy, distribution and evolution of the lentil and its wild relatives. In: Witcombe, J.R., Erskine, W. (Eds.), Genetic Resources and Their Exploitation—Chickpeas, Faba Beans and Lentils. Advances in Agricultural Biotechnology, Springer, Dordrecht, Netherlands, pp. 187–203.

D'Hont, A., Paget-Goy, A., Escoute, J., Carreel, F., 2000. The interspecific genome structure of cultivated banana, *Musa* spp. revealed by genomic DNA in situ hybridization. Theor. Appl. Genet. 100 (2), 177–183.

Dadu, R.H., Ford, R., Sambasivam, P., Gupta, D., 2017. A novel *Lens orientalis* resistance source to the recently evolved highly aggressive Australian ascochyta lentis isolates. Front. Plant Sci. 8, 1–7.

Driscoll, C.J., Sears, E.R., 1965. Mapping of a wheat-rye translocation. Genetics 51 (3), 439.

Ersfeld, K., 2004. Fiber-FISH: fluorescence in situ hybridization on stretched DNA. Methods Mol. Biol. 270, 395–402.

Erskine, W., 1997. Lessons for breeders from land races of lentil. Euphytica 93 (1), 107–112.

Erskine, W., Chandra, S., Chaudhry, M., Malik, I.A., Sarker, A., Sharma, B., Tufail, M., Tyagi, M.C., 1998. A bottleneck in lentil: widening its genetic base in South Asia. Euphytica 101 (2), 207–211.

Fernández, M., Ruiz, M.L., Linares, C., Fominaya, A., Pérez de la Vega, M., 2005. 5S rDNA genome regions of *Lens* species. Genome 48 (5), 937–942.

Fiala, J.V., Tullu, A., Banniza, S., Séguin-Swartz, G., Vandenberg, A., 2009. Interspecies transfer of resistance to anthracnose in lentil (Medic.). Crop Sci. 49 (3), 825–830.

Findley, S.D., Cannon, S., Varala, K., Du, J., Ma, J., Hudson, M.E., Birchler, J.A., Stacey, G., 2010. A fluorescence in situ hybridization system for karyotyping soybean. Genetics 185 (3), 727–744.

Findley, S.D., Birchler, J.A., Stacey, G., 2017. Metaphase chromosome preparation from soybean (*Glycine max*) root tips. Curr. Protoc. Plant Biol. 2 (1), 78–88.

Ford, R., Pang, E.C.K., Taylor, P.W.J., 1997. Diversity analysis and species identification in *Lens* using PCR generated markers. Euphytica 96 (2), 247–255.

Fratini, R., Ruiz, M.L., 2006. Interspecific hybridization in the genus *Lens* applying *in vitro* embryo rescue. Euphytica 150 (1), 271–280.

Fratini, R., Ruiz, M.L., 2011. Wide crossing in lentil through embryo rescue. Plant Embryo Cult.: Meth. Protoc. 710, 131–139.

Fratini, R., Ruiz, M.L., de la Vega, M.P., 2004. Intra-specific and inter-sub-specific crossing in lentil (*Lens culinaris* Medik.). Can. J. Plant Sci. 84 (4), 981–986.

Fredua-Agyeman, R., Coriton, O., Huteau, V., Parkin, I.A., Chèvre, A.M., Rahman, H., 2014. Molecular cytogenetic identification of B genome chromosomes linked to blackleg disease resistance in *Brassica napus* × *B. carinata* interspecific hybrids. Theor. Appl. Genet. 127 (6), 1305–1318.

Furuta, 1975. Quantitative variation of nuclear DNA in genus *Aegilops*. Jpn. J. Genet. 50 (5), 383–392.

Gaffarzadeh-Namazi, L., Asghari-Zakaria, R., Babaeian, N., Kazemi-Tabar, K., 2007. Comparative study of chromosome morphology and C-banding patterns in several genotypes of *Lens culinaris*. Pak. J. Biol. Sci. 10 (11), 1811–1816.

Galasso, I., 2003. Distribution of highly repeated DNA sequences in species of the genus *Lens* Miller. Genome 46 (6), 1118–1124.

Galasso, I., Schmidt, T., Pignone, D., 2001. Identification of *Lens culinaris* ssp. *culinaris* chromosomes by physical mapping of repetitive DNA sequences. Chromosom. Res. 9 (3), 199–209.

Gall, J.G., Pardue, M.L., 1969. Formation and detection of RNA-DNA hybrid molecules in cytological preparations. Proc. Natl. Acad. Sci. 63 (2), 378–383.

Gatti, I., Guindón, F., Bermejo, C., Espósito, A., Cointry, E., 2016. *In vitro* tissue culture in breeding programs of leguminous pulses: use and current status. Plant Cell Tissue Organ Cult. 127 (3), 543–559.

Gill, B.S., Burnham, C.R., Stringam, G.R., Stout, J.T., Winheimer, W.H., 1980. Cytogenetic analysis of chromosomal translocation in tomato: preferential breakage in heterochromatin. Can. J. Genet. Cytol. 22, 333–341.

Goshen, D., Ladizinsky, G., Muehlbauer, F.J., 1982. Restoration of meiotic regularity and fertility among derivatives of *Lens culinaris* x *L. nigricans* hybrids. Euphytica 31 (3), 795–799.

Goyal, S., Verma, R.C., 2015. Gamma ray and ethyl methane sulphonate induced translocation and inversion heterozygote in *Lens culinaris* Medik (lentil). Cytologia 80 (2), 231–235.

Greilhuber, J., Ebert, I., 1994. Genome size variation in *Pisum sativum*. Genome 37 (4), 646–655.

Guha, S., Maheshwari, S.C., 1964. *In vitro* production of embryos from anthers of *Datura*. Nature 204, 497.

Guha, S., Maheshwari, S.C., 1966. Cell division and differentiation of embryos in the pollen grains of *Daturain vitro*. Nature 212, 97–98.

Gupta, P.K., 1995. Cytogenetics. Rastogi and Company, Meerut, India.

Gupta, D., 2003. Genetic and Cytogenetic Studies in Populations Derived From Crosses Between Wild *Lens* Species and Cultivated Lentils (Doctoral dissertation). CSK Himachal Pradesh, Krishi Vishvavidyalaya, Palampur, India.

Gupta, P.K., Gupta, S.N., 1991. Cytogenetics of chromosome interchanges in plants. Chromosome Eng. Plants: Genet. Breed. Evol. 2A, 87–112.

Gupta, D., Sharma, S.K., 2005. Embryo-ovule rescue technique for overcoming post-fertilization barriers in interspecific crosses of *Lens*. J. Lentil Res. 2, 27–30.

Gupta, D., Sharma, S.K., 2006. Evaluation of wild *Lens* taxa for agro-morphological traits, fungal diseases and moisture stress in North Western Indian Hills. Genet. Resour. Crop. Evol. 53 (6), 1233–1241.

Gupta, D., Sharma, S.K., 2007. Widening the gene pool of cultivated lentils through introgression of alien chromatin from wild *Lens* subspecies. Plant Breed. 126 (1), 58–61.

Gupta, P.K., Singh, J., 1981. Standard karyotype in lentil var. Pant L-639. LENS (Lentil Exp. News Serv.) 8, 23.

Gupta, P.K., Kumar, S., Tyagi, B.S., Sharma, S.K., 1999. Chromosome interchanges in lentil (*Lens-culinaris* Med.). Cytologia 64, 387–394.

Gupta, M., Verma, B., Kumar, N., Chahota, R.K., Rathour, R., Sharma, S.K., Bhatia, S., Sharma, T.R., 2012. Construction of intersubspecific molecular genetic map of lentil based on ISSR, RAPD and SSR markers. J. Genet. 91, 279–287.

Hamdi, A., Erskine, W., 1994. Germplasm Program Legumes–Annual Report. ICARDA, Aleppo, Syria.

Hanson, R.E., Zwick, M.S., Choi, S., 1995. Fluorescent in situ hybridization of a bacterial artificial chromosome. Genome 38, 646–651.

Heitz, E., 1926. Der nachweis der chromosomen. Vergleichende studien uber ihre zahl, grosse und form im Pflanzenreich I. Zeitsch. Bot. 18, 625–681.

Hizume, M., Shibata, F., Matsusaki, Y., Garajova, Z., 2002. Chromosome identification and comparative karyotypic analyses of four Pinus species. Theor. Appl. Genet. 105 (4), 491–497.

Jha, T.B., Mahanti, A., Ghorai, A., 2015. Karyotype analysis of Indian lentils through EMA based Geimsa staining. Caryologia 68 (4), 280–288.

Jiang, J., Gill, B.S., 1994. New 18S· 26S ribosomal RNA gene loci: chromosomal landmarks for the evolution of polyploid wheats. Chromosoma 103 (3), 179–185.

Kannan, T.P., Zilfalil, B.A., 2009. Cytogenetics: past, present and future. Malaysian J. Med. Sci. 16 (2), 4.

Kasha, K.J., Kao, K.N., 1970. High frequency haploid production in barley (*Hordeum vulgare* L.). Nature 225 (5235), 874–876.

Kaur, P., Bhalla, J.K., 1998. Regeneration of haploid plants from microspore culture of pigeon pea (*Cajanus cajan* L.). Indian J. Exp. Bot. 36, 736–738.

Keller, W.A., Ferrie, A.M., 2002. Development of haploid production technology applicable for genetic improvement of lentils. In: Saskatchewan Agriculture & Food, Agriculture Development Fund (ADF) Project No. 98000297.

Keshwani, R., 1973. Cytological studies in some varieties of *Lens*. Proc. Indian Sci. Cong. Assoc. 60, 320.

Khattak, G.S.S., Wolny, E., Saeed, I., 2007. Detection of ribosomal DNA sites in chickpea (*Cicer arietinum* L.) and mungbean (*Vigna radiata* (L.) Wiltzek) by fluorescence in situ hybridization. Pak. J. Bot. 39, 1511–1515.

Khawar, K.M., Sancak, C., Uranbey, S., Özcan, S., 2004. Effect of thidiazuron on shoot regeneration from different explants of lentil (*Lens culinaris* Medik.) via organogenesis. Turk. J. Bot. 28 (4), 421–426.

Khentry, A.Y., Wang, A.S.H., Ford, R., 2014. *In vitro* propagation of six parental lentil (*Lens culinaris*ssp. *culinaris*) genotypes. US Open Agric. J. 1 (1), 1–8.

King, K., Torres, R.A., Zentgraf, U., Hemleben, V., 1993. Molecular evolution of the intergenic spacer in the nuclear ribosomal RNA genes of Cucurbitaceae. J. Mol. Evol. 36 (2), 144–152.

Koo, D.H., Jiang, J.M., 2009. Super-stretched pachytene chromosomes for fluorescence in situ hybridization mapping and immune-detection of DNA methylation. Plant J. 59, 509–516.

Kumar, S., Gupta, P.K., 1997. Pachytene chromosomes in lentil. Lentil Exp. News Serv. 24 (1–2), 31–34.

Kumar, S., Gupta, P.K., 1998. Aneuploids in the progeny of translocation heterozygotes in lentil (*Lens culinaris* Med.). Lens Newsl. 25 (1–2), 10–14.

Kumar, S., Ramesh, B., 2001. A study of nucleolar organizers in lentil using FISH and spore quartet analyses. Cytologia 66 (3), 247–252.

Ladizinsky, G., 1979. The origin of lentil and its wild gene pool. Euphytica 28 (1), 179–187.

Ladizinsky, G., Abbo, S., 1993. Cryptic speciation in *Lens culinaris*. Genet. Resour. Crop. Evol. 40 (1), 1–5.

Ladizinsky, G., Braun, D., Goshen, D., Muehlbauer, F.J., 1984. The biological species of the genus *Lens* L. Bot. Gaz. 145 (2), 253–261.

Ladizinsky, G., Cohen, D., Muehlbauer, F.J., 1985. Hybridization in the genus *Lens* by means of embryo culture. Theor. Appl. Genet. 70, 97–101.

Laurie, D.A., Bennett, M.D., 1985. Nuclear DNA content in the genera *Zea* and *Sorghum*. Intergeneric, interspecific and intraspecific variation. Heredity 55 (3), 307–313.

Lavania, U.C., Lavania, S., 1983. Karyotype studies in Indian pulses. Genet. Agrarian 37 (3–4), 279–307.

Le, H.T., Armstrong, K.C., Miki, B., 1989. Detection of rye DNA in wheat-rye hybrids and wheat translocation stocks using total genomic DNA as a probe. Plant Mol. Biol. Report. 7 (2), 150–158.

Leitch, I.J., Heslop-Harrison, J.S., 1992. Physical mapping of the 18S–5.8S–26S rRNA genes in barley by in situ hybridization. Genome 35 (6), 1013–1018.

Lombello, R.A., Pinto-Maglio, C.A., 2004. Heterochromatin and rDNA sites in *Coffea* L. chromosomes revealed by FISH and CMA/DAPI II: *C. canephora* cv. *Apoatã*, *C. salvatrix* and *C. sessiliflora*. Caryologia 57 (2), 138–143.

Mahama, A.A., Deaderick, L.M., Sadanaga, K., Newhouse, K.E., Palmer, R.G., 1999. Cytogenetic analysis of translocations in soybean. J. Hered. 90 (6), 648–653.

Malaviya, D.R., Shukla, R.S., 1983. Anisoploidy in *Lens culinaris*, occurrence and elimination from the natural population. Lentil Exp. News Serv. 10 (1), 19–20.

Malaviya, D.R., Shukla, R.S., 1984. Cytological instability of colchiploids in *Lens culinaris*. Lentil Exp. News Serv. 11 (2), 10–12.

Matsui, S.I., Sasaki, M., 1973. Differential staining of nucleolus organisers in mammalian chromosomes. Nature 246 (5429), 148–150.

Mayer, M.S., Soltis, P.S., 1994. Chloroplast DNA phylogeny of *Lens* (Leguminosae): origin and diversity of the cultivated lentil. Theor. Appl. Genet. 87 (7), 773–781.

Mehra, R.C., Butler, M.G., Beckman, T., 1986. N banding and karyotype analysis of *Lens culinaris*. J. Hered. 77 (6), 473–474.

Micke, A., Donini, B., Maluszyński, M., 1990. Induced mutations for crop improvement. In: Mutation Breeding Review. vol. 7. IAEA, Vienna, pp. 1–41.

Miranda, F., 1931. Sobre las algas y cianofíceas del Cantábrico, especialmente de Gijón. Trab. Mus. Nac. Cienc. Nat. Ser. Bot. Madrid 25, 1–106.

Mix, G., Wang, H.M., 1988. *In vitro* generation of haploid plants (*Vigna unguiculata* L.). Landbauforsch. Volk. 38, 305–309.

Mohamed, M.F., Read, P.E., Coyne, D.P., 1992. Dark preconditioning, CPPU, and thidiazuron promote shoot organogenesis on seedling node explants of common and faba beans. J. Am. Soc. Hortic. Sci. 117 (4), 668–672.

Mukai, Y., 1996. Multicolor fluorescence in situ hybridization: a new tool for genome analysis. In: Jauhar, P.P. (Ed.), Methods of Genome Analysis in Plants. CRC Press, Boca Raton, pp. 181–192.

Mukai, Y., Endo, T.R., Gill, B.S., 1991. Physical mapping of the 18S. 26S rRNA multigene family in common wheat: identification of a new locus. Chromosoma 100 (2), 71–78.

Nagar, A., 2013. Physical and chemical mutagens induced the mutation in M3 generation of Green gram (*Vigna radiata* L. Wilczek). Int. J. Curr. Sci. 6, 58–62.

Naithani, S.P., Sarbhoy, R.K., 1973. Cytological studies in *Lens esculenta* Moench. Cytologia 38 (2), 195–203.

Nandanwar, R.S., Narkhede, M.M., 1991. Intraspecific variation in karyotype of lentil. J. Maharashtra Agric. Univ. 16, 24–27.

Pal, T., Ghosh, S., Mondal, A., De, K.K., 2016. Evaluation of genetic diversity in some promising varieties of lentil using karyological characters and protein profiling. J. Genet. Eng. Biotechnol. 14 (1), 39–48.

Pardue, M.L., Gall, J.G., 1970. Chromosomal localization of mouse satellite DNA. Science 168 (3937), 1356–1358.

Pedersen, C., Linde-Laursen, I.B., 1994. Chromosomal locations of four minor rDNA loci and a marker microsatellite sequence in barley. Chromosom. Res. 2 (1), 65–71.

Pinkel, D., Straume, T., Gray, J.W., 1986. Cytogenetic analysis using quantitative, high-sensitivity, fluorescence hybridization. Proc. Natl. Acad. Sci. 83 (9), 2934–2938.

Pires, A., Geraldo, N., Cerqueira, T., Abranches, R., 2012. Integrated approaches to studying *Medicago truncatula* genome structure and function and their applications in biotechnology. Mol. Breed. 30 (3), 1431.

Podder, R., Banniza, S., Vandenberg, A., 2013. Screening of wild and cultivated lentil germplasm for resistance to stemphylium blight. Plant Genet. Resour. 11 (1), 26–35.

Pratap, A., Sethi, G.S., Chaudhary, H.K., 2005. Relative efficiency of different Gramineae genera for haploid induction in triticale and triticale × wheat hybrids through the chromosome elimination technique. Plant Breed. 124, 147–153.

Pratap, A., Choudhary, A.K., Kumar, J., 2010. *In vitro* techniques towards genetic enhancement of food legumes—a review. J. Food Legum. 23, 169–185.

Pundir, R.P.S., Rao, N.K., Van Der Maesen, L.J.G., 1983. Induced autotetraploidy in chickpea (*Cicer arietinum* L.). Theor. Appl. Genet. 65 (2), 119–122.

Ramesh, B., 1994. Karyotype, nuclear and chromosomal DNA variation in *Lens culinaris* Med. Cytologia 59 (1), 7–15.

Ramzan, F., Younis, A., Lim, K.B., 2017. Application of genomic in situ hybridization in horticultural science. Int. J. Genom. 2017, 1–12.

Rayburn, A.L., Gill, B.S., 1985. Use of biotin-labeled probes to map specific DNA sequences on wheat chromosomes. J. Hered. 76 (2), 78–81.

Raziuddin, A.H., Altaf, C.M., Hanan, F., 1990. Investigation on the karyotype of lentil (*Lens culinaris* Medik). Sarhad J. Agric. 6, 261–264.

Rees, T.A.V., 1984. Sodium dependent photosynthetic oxygen evolution in a marine diatom. J. Exp. Bot. 35 (3), 332–337.

Renfrew, J.M., 1969. The archaeological evidence for the domestication of plants: methods and problems. In: Peter, J.U., Dimbleby, G.W. (Eds.), The Domestication and Exploitation of Plants and Animals. Aldine Publishing Company, Chicago, pp. 149–172.

Robledo, G., Lavia, G.I., Seijo, G., 2009. Species relations among wild *Arachis* species with the A genome as revealed by FISH mapping of rDNA loci and heterochromatin detection. Theor. Appl. Genet. 118 (7), 1295–1307.

Rowley, J.D., 1973. A new consistent chromosomal abnormality in chronic myelogenous leukaemia identified by quinacrine fluorescence and Giemsa staining. Nature 243 (5405), 290–293.

Saha, S., Tullu, A., Yuan, H.Y., Lulsdorf, M.M., Vandenberg, A., 2015. Improvement of embryo rescue technique using 4-chloroindole-3 acetic acid in combination with in vivo grafting to overcome barriers in lentil interspecific crosses. Plant Cell Tissue Organ Cult. (PCTOC) 120 (1), 109–116.

Sarker, R.H., Biswas, A., Mustafa, B.M., Mahbub, S., Hoque, M.I., 2003. Agrobacterium-mediated transformation of lentil (*Lens culinaris* Medik.). Plant Tissue Cult. 13, 1–12.

Schweizer, D., Davis, D.R., 1972. Nuclear DNA content of *Pisum* genotypes grown *in vivo* and *in vitro*. *Planta* (Berl.) 106, 23–29.

Seijo, J.G., Fernández, A., 2003. Karyotype analysis and chromosome evolution in South American species of *Lathyrus* (Leguminosae). Am. J. Bot. 90 (7), 980–987.

Sharma, A.K., 1962. Karyotype constancy in different strains of *Lensesculenta* Moench.as worked out through recent techniques. Indian Agric. 7, 103–111.

Sharma, C.B.S.R., 1990. Chemically induced aneuploidy in higher plants. Mutagenesis 5 (2), 105–126.

Sharma, P.C., Gupta, P.K., 1982. Karyotypes in some pulse crops. Nucleus 25 (3), 81–85.

Sharma, D.R., Kaur, R., Kumar, K., 1996. Embryo rescue in plants—a review. Euphytica 89 (3), 325–337.

Sharma, S.K., Yamamoto, M., Mukai, Y., 2016. Molecular cytogenetic approaches in exploration of important chromosomal landmarks in plants. In: Molecular Breeding for Sustainable Crop Improvement. Springer International Publishing, pp. 127–148.

Singh, R.J., 2003. Plant cytogenetics, second ed. CRC Press, Boca Raton, FL, USA.

Singh, M., Rana, M.K., Kumar, K., Bisht, I.S., Dutta, M., Bansal, K.C., 2013. Broadening the genetic base of lentil cultivars through inter-sub-specific and interspecific crosses of *Lens* taxa. Plant Breed. 132, 667–675.

Singh, M., Rana, J.C., Singh, B., Kumar, S., Saxena, D.R., Saxena, A., Rizvi, A.H., Sarker, A., 2017. Comparative agronomic performance and reaction to fusarium wilt of *Lens culinaris* × *L. orientalis* and *L. culinaris* × *L. ervoides* derivatives. Front. Plant Sci. 8, 1–12. https://doi.org/10.3389/fpls.2017.01162.

Sinha, S.S.N., Acharia, S.S., 1972. Karyotype analysis in some varieties of *Lens culinaris*. Cytologia 37 (4), 673–683.

Slinkard, A.E., 1985. Cytology and cytogenetics of lentils (*Lens culinaris*). Lens Newsl. (ICARDA) 12 (2), 1–10.

Sonnante, G., Galasso, I., Pignone, D., 2003. ITS sequence analysis and phylogenetic inference in the genus *Lens* mill. Ann. Bot. 91 (1), 49–54.

Srivastava, R.P., Vasishtha, H., 2012. Saponins and lectins of Indian chickpeas (*Cicer arietinum*) and lentils (*Lens culinaris*). Indian J. Agric. Biochem. 25 (1), 44–47.

Suvorova, G., 2014. Hybridization of cultivated lentil (*Lens culinaris* Medik.) and wild species *Lens tomentosus* Ladizinsky. Czech J. Genet. Plant Breed. 50 (2), 130–134.

Tavallaie, F.Z., Ghareyazie, B., Bagheri, A., Sharma, K.K., 2011. Lentil regeneration from cotyledon explant bearing a small part of the embryo axis. Plant Tissue Cult. Biotechnol. 21 (2), 169–180.

Tullu, A., Diederichsen, A., Suvorova, G., Vandenberg, A., 2011. Genetic and genomic resources of lentil: status, use and prospects. Plant Genet. Resour. 9 (1), 19–29.

Tullu, A., Bett, K., Banniza, S., Vail, S., Vandenberg, A., 2013. Widening the genetic base of cultivated lentil through hybridization of *Lensculinaris* 'Eston' and *L. ervoides* accession IG 72815. Can. J. Plant Sci. 93 (6), 1037–1047.

Vandenberg, A., Slinkard, A.E., 1989. Inheritance of four new qualitative genes in lentil. J. Hered. 80 (4), 320–322.

Williams, J.T., Sanchez, A.M.C., Jackson, M.T., 1974. Studies on lentils and their variation: 1. The taxonomy of the species. SABRAO J. Breed. Genet. 6 (2), 133–145.

Wong, M.M., Gujaria-Verma, N., Ramsay, L., Yuan, H.Y., Caron, C., Diapari, M., Vandenberg, A., Bett, K.E., 2015. Classification and characterization of species within the genus *Lens* using genotyping-by-sequencing (GBS). PLoS ONE 10 (3), e0122025.

Ye, G., McNeil, D.L., Conner, A.J., Hill, G.D., 2002. Multiple shoot formation in lentil (*Lens culinaris*) seeds. N. Z. J. Crop. Hortic. Sci. 30 (1), 1–8.

Yuan, H.Y., Lulsdorf, M., Tullu, A., Gurusamy, V., Vandenberg, A., 2011. In vivo grafting of wild *Lens* species to *Vicia faba* rootstocks. Plant Genet. Resour. 9 (4), 543–548.

Zhang, A., Li, W., Wang, C., Yang, X., Chen, C., Zhu, C., Peng, N., Tian, Z., Wang, Y., Zhang, H., Liu, X., 2017. Molecular cytogenetics identification of a wheat–*Leymus mollis* double disomic addition line with stripe rust resistance. Genome 60 (5), 375–383.

Zhao, X.C., Sharp, P.J., 1998. Production of all eight genotypes of null alleles at 'waxy'loci in bread wheat, *Triticum aestivum* L. Plant Breed. 117 (5), 488–490.

Zohary, D., 1972. The wild progenitor and the place of origin of the cultivated lentil: *Lens culinaris*. Econ. Bot. 26 (4), 326–332.

FURTHER READING

Gustavsson, I., 1969. Cytogenetics, distribution and phenotypic effects of a translocation in Swedish cattle. Hereditas 63 (1-2), 68–169.

Jewell, D.C., 1981. Recognition of two types of positive staining chromosomal material by manipulation of critical steps in the N-banding technique. Stain. Technol. 56 (4), 227–234.

Staginnus, C., Desel, C., Schmidt, T., Kahl, G., 2010. Assembling a puzzle of dispersed retrotransposable sequences in the genome of chickpea (*Cicer arietinum* L.). Genome 53 (12), 1090–1102.

Chapter 6

Gene Pyramiding and Multiple Character Breeding

Maneet Rana*, Ankita Sood†, Waseem Hussain‡, Rahul Kaldate§, Tilak Raj Sharma¶, R.K. Gill†, Shiv Kumar‖, Sarvjeet Singh†

**Division of Crop Improvement, ICAR-Indian Grassland and Fodder Research Institute, Jhansi, India, †Department of Plant Breeding and Genetics, Punjab Agricultural University, Ludhiana, India, ‡Department of Agronomy and Horticulture, University of Nebraska, Lincoln, NE, United States, §Department of Agricultural Biotechnology, CSKHPKV, Palampur, India, ¶ICAR-Indian Institute of Agricultural Biotechnology, Ranchi, India, ‖Biodiversity and Integrated Gene Management Program, International Center for Agricultural Research in the Dry Areas (ICARDA), Rabat, Morocco*

6.1. INTRODUCTION

Lentil (*Lens culinaris* ssp. *culinaris* Medik.) is an autogamous diploid ($2n = 2x = 14$) species, likely to have originated in southern Turkey/northern Syria and later to have spread to the Mediterranean Basin, Central Asia, Nile Delta, and the Indian subcontinent (Sandhu and Singh, 2007; Cubero et al., 2009). It is the fourth most important cool season grain legume in the world. Globally, lentils are cultivated on 4.52 million hectares with annual production of 4.82 million tons in as many as 52 countries (FAOSTAT, 2016). The major lentil-producing countries are Canada, India, Turkey, Australia, Nepal, Bangladesh, United States, Ethiopia, China, and Iran.

Cubero et al. (2009) reviewed lentil phylogeny, origin, domestication, and spread. The most commonly accepted classification is the one proposed by Ferguson et al. (2000), using morphology and molecular markers. The classification of the genus *Lens* as proposed by Ferguson et al. (2000) comprises: *L. culinaris* (with four subspecies, i.e., *culinaris, orientalis, tomentosus*, and *odemensis*), *L. ervoides*, *L. nigricans,* and *L. lamottei.* The genus now consists of seven taxa split into four species:

1. *L. culinaris* Medikus
   a. ssp. culinaris
   b. ssp. orientalis (Boiss.) Ponert
   c. ssp. tomentosus (Ladiz.) M.E. Ferguson et al.
   d. ssp. odemensis (Ladiz.) M. E. Ferguson et al.

Lentils. https://doi.org/10.1016/B978-0-12-813522-8.00006-6

2. L. ervoides (Brign.) Grande
3. *L. nigricans* (M. Bieb.) Godr.
4. *L. lamottei* Czefr.

Lentils are well adapted to dry areas with <400 mm rainfall. It provides important dietary compounds like protein, carbohydrates, fiber, minerals, vitamins, and antioxidants. With 26%–30% protein, lentils have the third highest level of protein from any plant-based food after soybeans and hemp, and is an important part of the diet in many parts of the world, especially in the Indian subcontinent, which has a large vegetarian population. Lentils are known for their high nutritive value, being rich in protein, prebiotics, and micronutrients including iron, zinc, and β-carotene (Erskine and Sarker, 2004).

Lentils help in the management of Type 2 diabetes due to their low glycemic index (<55), thus their effect on blood glucose level is less in comparison to other foods containing carbohydrates. Its straw, including husks of lentils along with its stem, dried leaves, and bran residues, is a valued livestock animal feed consisting of minerals and carbohydrates, at 2% and 59%, respectively (Frederick et al., 2006). In addition, its seeds are a source of commercial starch which is used in industry for printing and textiles (Kay, 1979). It is an important cool season food legume crop generally grown in rotation with cereals to break cereal disease cycles and to fix atmospheric nitrogen, thus reducing the demand for nitrogen fertilizers (Anjam et al., 2005). So far, classical plant breeding approaches utilizing selection-recombination and selection cycles have contributed successfully to improve lentil crops. These approaches are imprecise and time-consuming, particularly for improving complex quantitative traits. The recent developments in molecular marker technologies have made it possible to localize genomic regions and assess their phenotypic effects on various quantitative traits.

Molecular markers play an important role in improving our understanding of the genetic basis of economically important traits and are efficient tools to speed up crop improvement (Varshney and Tuberosa, 2007). They can be efficiently used for selection of traits with low heritability, gene introgression coming from native or exotic germplasm (Gupta et al., 1996), estimation of genetic relatedness among accessions, cultivar description (Smith et al., 1992), and the identification of quantitative trait loci (QTLs) that control important agronomic traits (Dudley, 1993). They offer plant breeders a rapid and precise alternative approach to conventional selection schemes for improving cultivars for quantitative traits like yield, adaptability, and resistance to biotic and abiotic stresses, etc.

With the availability of polymorphic markers and linkage maps, QTLs for various traits including plant height, days to flowering, winter hardiness, pod dehiscence, and growth habits in lentils have been identified and mapped, using both inter- and intraspecific maps (Tarán et al., 2003; Kahraman et al., 2004;

Fratini et al., 2007; Tullu et al., 2008). Similarly, QTLs have been identified for resistance to diseases like ascochyta blight, anthracnose, rust, and stemphylium blight (Rubeena et al., 2006; Saha et al., 2010b; Fikru et al., 2014). Studies on identifying the QTLs linked to yield-related traits are limited in lentils. For seed weight in lentil, QTLs have been located by Abbo et al. (1991). Fratini et al. (2007) mapped QTLs for seed diameter and seed weight in lentils. Recently, QTLs for seed-related traits, such as seed size and shape, seed weight, seed diameter, seed thickness, seed plumpness have also been identified in lentils (Fedoruk et al., 2013; Verma et al., 2015).

Molecular markers linked to the major genes and QTLs in combination with linkage maps and genomics can be used to aid in selecting and improving the plants. Molecular breeding strategies can also be used to integrate multiple traits into a single genetic background. The various molecular breeding strategies to transfer or introgress multiple traits or genes include marker-assisted selection (MAS), marker-assisted backcrossing (MABC), marker-assisted gene pyramiding, marker-assisted recurrent selection (MARS), and genome-wide selection (GWS), and genomic selection (GS). In this chapter, we address conventional and molecular breeding strategies to integrate multiple traits or genes into a single genetic background. We begin with the important target traits in lentils, the importance of the genetic association between the traits, breeding schemes to integrate multiple traits, and principles and procedures of various molecular breeding approaches as powerful tools to breed multiple traits in lentil. A general overview of various breeding methodologies, including conventional and molecular, to breed for multiple traits and genes is given in Fig. 6.1, and Table 6.1 lists region-wise targeted traits in lentils for improvement across the globe.

6.2. CONVENTIONAL BREEDING APPROACHES TO INTEGRATE MULTIPLE TRAITS

6.2.1 Important Traits to be Focused Upon

The first challenge in breeding for multiple traits is to determine and prioritize traits which are most important for the target environment and market. Several traits of importance can simultaneously be targeted for genetic improvement of lentil cultivars. However, prioritization of traits is very important, as there is a cost for every trait the plants express in the final phenotype. Breeders should focus on the identification of genotypes with desired adaptation to biotic and abiotic stresses, superior grain quality, nutritional attributes, and appropriate phenology to match with the environment. In general, traits that can be used for genetic improvement of lentil are summarized in Fig. 6.2. For multiple trait selections and integration, breeders should focus on the traits that are associated genetically.

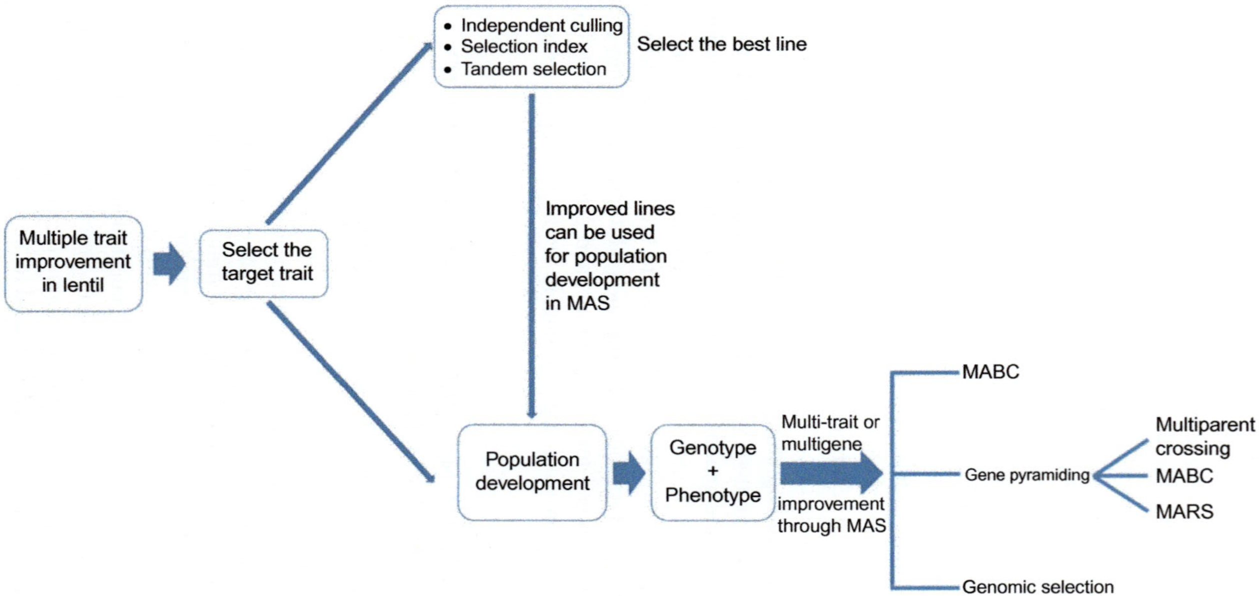

FIG. 6.1 The general overview of various breeding methodologies including conventional and molecular to breed for multiple traits and genes.

TABLE 6.1 List of Region-Wise Targeted Traits in Lentils for Improvement Across the Globe

S. No.	Region	Targeted Trait
1.	Africa	Heat and drought tolerance, Ascochyta blight resistance, Anthracnose resistance, rust resistance, nutritional enhancement, yield and yield-related traits
2.	Eastern Europe	Heat and drought tolerance, cold tolerance, Ascochyta blight resistance, *Fusarium* wilt resistance, Anthracnose resistance, nutritional enhancement, yield and yield-related traits
3.	North Africa	Heat and drought tolerance, cold tolerance, salinity tolerance, Ascochyta blight resistance, Fusarium wilt resistance, Anthracnose resistance, Botrytis gray mold resistance, rust resistance, Stemphylium blight, Nutritional enhancement, yield and yield-related traits
4.	North America	Heat and drought tolerance, cold tolerance, salinity tolerance, Ascochyta blight resistance, Anthracnose resistance, Botrytis gray mold resistance, rust resistance, Stemphylium blight, nutritional enhancement, yield and yield-related traits
5.	Oceania	Heat and drought tolerance, boron tolerance, Ascochyta blight resistance, Botrytis gray mold resistance, nutritional enhancement, yield and yield-related traits
6.	Russia	Heat and drought tolerance, Ascochyta blight resistance, nutritional enhancement, yield and yield-related traits
7.	South America	Heat and drought tolerance, cold tolerance, Ascochyta blight resistance, Fusarium wilt resistance, Botrytis gray mold resistance, rust resistance, nutritional enhancement, yield and yield-related traits
8.	South Asia	Heat and drought tolerance, cold tolerance, salinity tolerance, boron tolerance, Ascochyta blight resistance, *Fusarium* wilt resistance, Anthracnose resistance, Botrytis gray mold resistance, rust resistance, Stemphylium blight, nutritional enhancement, yield and yield-related traits
9.	Western Asia	Heat and drought tolerance, cold tolerance, salinity tolerance, boron tolerance, Ascochyta blight resistance, Fusarium wilt resistance, nutritional enhancement, yield and yield-related traits
10.	Western Europe	Heat and drought tolerance, Ascochyta blight resistance, *Fusarium* wilt resistance, Botrytis gray mold resistance, rust resistance, nutritional enhancement, yield and yield-related traits

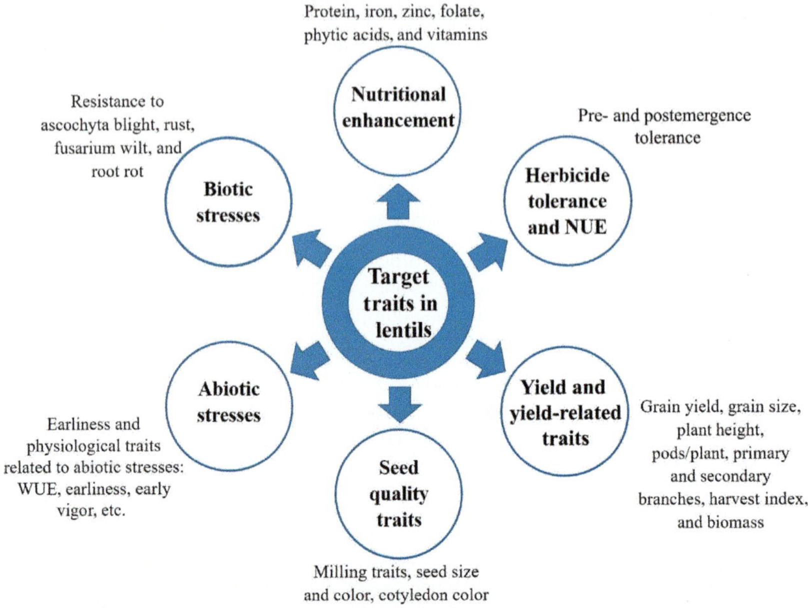

FIG. 6.2 Target lentil traits for potential genetic improvement.

6.2.2 Knowledge of Genetic Association Between Target Traits

In lentils or any other crop, breeders need to integrate several traits into a single variety for it to be successful and valued by farmers, markets, and consumers. The challenge for a breeder to breed for multiple traits is how to select simultaneously for multiple traits. The selection for multiple traits largely depends upon the degree of variability and association among the traits under consideration. Any pair of traits under selection may be associated (favorably or unfavorably) or independent. The association between the two traits or more may be genetic (includes both additive and dominance effects) or nongenetic (environmental). Genetic correlations between two traits that include additive effects, also called additive correlation, can be estimated in a breeding population using mating designs (e.g., diallel, nested, or factorial mating designs) and progeny testing of the genotypes in multiple environments. Additive genetic correlation is important in selection programs as it determines the degree of association between the traits by way of breeding values of individuals or additive effects. In other words, additive correlation indicates the extent to which selection for one trait will result in an indirect response to selection for the second trait (Bernardo, 2001). This indirect change is known as correlated response to selection and largely depends upon the degree of association between the two traits under selection. Traits in lentils that exhibit favorable correlated additive genetic effects provide the breeder with a bonus, as selection for one trait will cause an indirect change in the mean of the second trait.

Genetic correlation may be attributed to linkage disequilibrium and or pleiotropism. Genetic correlations due to linkage disequilibrium are unstable and random mating of individuals can break these linkages. However, genetic correlations due to pleiotropy (same loci controlling different traits) are stable. Depending upon the extent and nature of genetic correlations, correlated response to selection can be desirable or undesirable. Traits with unfavorable associations will be of concern to the breeder if the cause is unfavorably correlated genetic effects, especially those resulting from pleiotropy, and will pose a linkage drag (Luby and Shaw, 2009). Further, for a successful correlated response, the correlation among the traits should be near 1.0, and the heritability of the second trait should be near 0.9. Hence in lentil crops, the breeder should know the extent and nature of the association between the traits and heritability of traits to breed for multiple traits. Also, the breeder can obtain information from previous studies on heritability and correlation between the traits, types of gene action which may serve as a useful guide for the breeder's decisions to breed for multiple traits and genetic improvement of lentils. Information on the correlations between various traits in lentils is presented in Table 6.2.

6.2.3 Breeding Schemes for Multiple Traits

Improvement programs in lentils must consider more than one trait. Several methods have been developed to deal with the task of multiple trait selection and integration. To breed for multiple traits in lentils, breeders must determine the traits which are most important, keeping in mind the breeding populations and resources. Three basic strategies including tandem selection, independent culling, and selection index can be utilized to simultaneously breed or select for multiple traits.

i. Tandem selection

   Tandem selection attempts to improve a breeding population for several traits by selecting one trait at one time for several generations, then another trait is focused on for next breeding cycle or period (serial improvement). The major consideration for a breeder is to know how long each trait is selected for before switching to another trait and at what intensity the trait is selected. Tandem selection is effective when correlation does not exist between the traits or the relative importance of each trait changes throughout the years. For example, if genetic correlations do not exist between, for example, yield and disease resistance, tandem selection can be used effectively to increase the level of disease resistance before selection for yield is started.

ii. Independent culling

   Independent culling, also called truncation selection, involves selecting for multiple traits in one generation in a specified order from a single population. For example, in family selection, means from replicated trials from 400 half-sib families are to be used. First, the breeder can select the 160

TABLE 6.2 Correlation Among Various Traits in Lentil

S. No.	Trait	Positive	Negative	Level of Significance	Reference
1.	Grain yield/plant	Number of pods/plant, number of branches/plant, and number of grains/pod	100-grain weight	1%	Nandan and Pandya (1980)
	100-grain weight	Number of branches/plant	Number of pods/plant and number of grains/pod		
	Number of grains/pod	Number of branches/plant and number of pods/plant			
2.	Grain yield per plant	Days to flowering, plant height, branches/plant, biological yield/plant, harvest index, and 1000-grain weight		5%	Bakshs et al. (1993)
	1000-grain weight	Days to flowering, plant height, branches/plant, biological yield/plant and harvest index		1% and 5%	
	Biological yield /plant	Days to flowering and branches/plant	Plant height	5%	
3.	Seed yield	Plant height, pods/peduncle, pods/plant, biomass, and straw yield	Pod dehiscence and viral diseases	1% and 5%	Anjam et al. (2005)
	Straw yield	Plant height, pods/peduncle, viral diseases and biomass	Pods/plant and pod dehiscence	1%	
	Biomass	Plant height, pods/peduncle, and pods/plant	Pod dehiscence and viral diseases	5%	
4.	Seed yield	Number of pods/plant, 100-seed weight, and harvest index	Plant height and biomass yield	1% and 5%	Salehi et al. (2008)
	Harvest index		Plant height, number of pods/plant, 100-seed weight and biomass yield	1%	

5.	Grain yield	Days to maturity, plant height, first pod height, biological yield/plant, number of pods/plant, and number of seeds/plant	Days to 50% flowering and number of branches/plant	1% and 5%	Tuba and Sakar (2008)
	Number of seeds/plant	Plant height, first pod height, biological yield/plant, number of pods/plant, and number of branches/plant	Days to 50% flowering and days to maturity	1%	
	Number of pods/plant	Days to maturity, plant height, first pod height, biological yield/plant, and number of branches/plant	Days to 50% flowering	1% and 5%	
	First pod height	Days to maturity, plant height, and biological yield/plant	Days to 50% flowering	1%	
	Plant height	Days to maturity and biological yield/plant	Days to 50% flowering	1%	
6.	Seed yield/plot	Pods/plant, seeds/pod, biological yield/plot, and harvest index	Days to 50% flowering, days to maturity, plant height, fruiting branches/plant, and 100-seed weight	1% and 5%	Singh et al. (2009)
	Harvest index	Pods/plant, seeds/pod, and biological yield/plot	Days to 50% flowering, days to maturity, plant height, fruiting branches/plant, and 100-seed weight	1%	
	Biological yield/plot	Plant height, fruiting branches/plant, pods/plant, and seeds/pod	Days to 50% flowering, days to maturity, and 100-seed weight	1%	
	100-seed weight	Days to 50% flowering, days to maturity, plant height, fruiting branches/plant, and pods/plant	Seeds/pod	1%	
	Seeds/pod	Plant height and pods/plant	Days to 50% flowering, days to maturity, and fruiting branches/plant	1%	
	Pods/plant	Days to 50% flowering and fruiting branches/plant	Days to 50% flowering and days to maturity	1% and 5%	
	Plant height	Days to 50% flowering and days to maturity		1%	

Continued

TABLE 6.2 Correlation Among Various Traits in Lentil—cont'd

S. No.	Trait	Positive	Negative	Level of Significance	Reference
7.	Seed yield	Germination, plant height, pods/plant, biological yield, 100-seed weight, and harvest index	Days to 50% flowering, days to maturity, branches/plant, pod size and seeds/pod	1% and 5%	Tyagi and Khan (2010)
	Harvest index	Germination, days to 50% flowering, plant height, branches/plant, pods/plant, pod size, seeds/pod, and 100-seed weight Negative:	Days to maturity and biological yield	1%	
	100-seed weight	Germination, days to maturity, plant height, branches/plant, and pod size	Days to 50% flowering, pods/plant, seeds/pod, and biological yield	5%	
	Seeds/pod	Days to maturity, pods/plant, and pod size	Germination, days to 50% flowering, plant height, and branches/plant	5%	
	Pods/plant	Germination, days to maturity, and plant height	Days to 50% flowering and branches/plant	1%	
	Branches/plant	Germination, days to 50% flowering, and days to maturity	Plant height	1%	
	Plant height	Germination	Days to 50% flowering and days to maturity	5%	
8.	Seed weight	Starch, RFO, and sucrose	Amylose and protein	1% and 5%	Tahir et al. (2011)
	Starch	Amylose and protein		1% and 5%	
	Amylose		RFO	5%	

9.	Green percent	Grain yield and biomass	Days to flowering and harvest index	1% and 5%	Aghili et al. (2012)
	Days to flowering	Number of hooks, harvest index, and primary branches/plant		1% and 5%	
	Number of hooks	Hook size, maturity, plant height, lowest pod, number of filled pods/plant, seed number/100 pods, primary branches/plant, secondary branches, biomass, and 100-seed weight		1%	
	Hook size	Grain yield, maturity, plant height, lowest pod, number of filled pods/plant, seed number/100 pods, primary branches/plant, secondary branches, biomass, and 100-seed weight		1% and 5%	
	Grain yield	Plant height, number of filled pods/plant, seed number/100 pods, biomass, and 100-seed weight		1% and 5%	
	Maturity	Plant height, lowest pod, number of filled pods/plant, seed number/100 pods, primary branches/plant, secondary branches, biomass, and 100-seed weight		1%	
	Plant height	Lowest pod, number of filled pods/plant, seed number/100 pods, primary branches/plant, secondary branches, biomass, and 100-seed weight		1%	
	Lowest pod	Number of filled pods/plant, seed number/100 pods, primary branches/plant, secondary branches, biomass, and 100-seed weight		1% and 5%	
	Number of filled pods/plant	Seed number/100 pods, primary branches/plant, secondary branches, biomass, and 100-seed weight		1%	
	Seed number/100 pods	Primary branches/plant, secondary branches, biomass, and 100-seed weight		1%	
	Primary branches/plant	Secondary branches, biomass, and 100-seed weight		1% and 5%	
	Secondary branches	Biomass and 100-seed weight		1%	

Continued

TABLE 6.2 Correlation Among Various Traits in Lentil—cont'd

S. No.	Trait	Positive	Negative	Level of Significance	Reference
10.	Seed weight/plant	Number of branches/plant, number of pods/plant, number of seeds/plant, and 100-seed weight	Days to flowering and days to maturity	1%	Ashrie et al. (2012)
	100-seed weight	Number of pods/plant	Days to flowering and days to maturity	1% and 5%	
	Number of seeds/plant	Number of pods/plant and number of branches/plant	Days to flowering	1%	
	Number of pods/plant	Number of branches/plant	Days to flowering	1%	
11.	Seed yield/plant	First pod height, number of branches/plant, number of pods/plant, 1000-seed weight, and number of seed/pod	Plant height		Karadavut and Kavurmaci (2013)
	Plant height	First pod height and number of pods/plant	Number of branches/plant, 1000-seed weight, and number of seed/pod		
	First pod height	Number of branches/plant and number of pods/plant	1000-seed weight and number of seed/pod		
	Number of branches/plant	Number of pods/plant, 1000-seed weight, and number of seed/pod			
	Number of pods/plant		1000-seed weight and number of seed/pod		
12.	Number of secondary branches	Number of pods/plant		1%	Taiery and Mirshekari (2014)
	Number of pods/plant	Grain yield		5%	
	100-seed weight	Grain yield		1%	

13.	Seed yield/plant	Days to maturity, plant height, number of primary branches/plant, number of secondary branches/plant, total number of pods/plant, number of effective pods/plant, number of seeds/plant, and number of seeds/pod		1%	Pandey et al. (2015)
	100-seed weight		Days to maturity, plant height, number of secondary branches/plant, total number of pods/plant, number of effective pods/plant, number of seeds/plant, and number of seeds/pod	1% and 5%	
	Number of seeds/pod	Days to maturity, plant height, total number of pods/plant, number of effective pods/plant, and number of seeds/plant		1%	
	Number of seeds/plant	Days to maturity, plant height, number of primary branches/plant, number of secondary branches/plant, total number of pods/plant, and number of effective pods/plant		1%	
	Number of effective pods/plant	Days to maturity, plant height, number of primary branches/plant, number of secondary branches/plant, and total number of pods/plant		1%	
	Total number of pods/plant	Days to maturity, plant height, number of primary branches/plant, and number of secondary branches/plant		1%	
	Number of secondary branches/plant	Number of primary branches/plant		1%	
	Number of primary branches/plant	Plant height	Days to maturity	1% and 5%	

Continued

TABLE 6.2 Correlation Among Various Traits in Lentil—cont'd

S. No.	Trait	Positive	Negative	Level of Significance	Reference
14.	Pod length	Number of primary branches/plant, pods/plant, days to 50% flowering, days to maturity, seed yield/plant, and 100-seed weight		1% and 5%	Kumar et al. (2016)
	Seed yield/plant	Plant height, number of primary branches/plant, pods/plant, days to 50% flowering, and days to maturity		1%	
	Days to maturity	Plant height, number of primary branches/plant, pods/plant, and days to 50% flowering	Seeds/pod	1% and 5%	
	Days to 50% flowering	Number of primary branches/plant and pods/plant		1%	
	Number of seeds/pod		Plant height	5%	
	Number of pods/plant	Number of primary branches/plant		1%	

best families (40%) based on yield. From this sample of 160, a selection intensity of 50% (80 families) can be used for seed size, followed with 50% selection intensity for winter hardiness. The total selection intensity would be $0.40 \times 0.50 \times 0.50 = 0.10$ or 10%, and only the 40 best families would be used for recombination breeding. Before using independent culling, breeders must keep in mind the following points:

(a) Breeders must maintain a sufficiently large population after each culling level to ensure that sufficient variation remains for the traits in subsequent culling.

(b) Breeders must apply less strict culling for the first trait to ensure there is sufficient variation for an unfavorably correlated trait.

(c) When genetic correlations are unimportant, breeders must keep in mind the order of culling, depending upon the economics of the breeding program and importance of the trait. For example, culling for rust disease resistance might be done at the seedling stage in a greenhouse, and only genotypes not culled at the seedling stage could be further screened for yield and quality traits in the field later.

iii. Selection index

One of the major difficulties in selection for multiple traits is the negative correlation between desirable traits. To obtain desirable genotypes, it is important to overcome the negative correlation between the traits. Index selection is a tool for improving traits simultaneously when a negative correlation exists between the traits. In index selection, the breeder creates a single new trait, the "index," which is a function of the multiple traits that are under selection. Index selection is basically a method for weighting individual traits based on their economic importance and the opportunity for the improvement. Index selection is theoretically the most efficient method for improving crop merit, given the concept of aggregate genetic value (Bernardo, 2001). The aggregate genetic value of the genotype is its performance based on multiple traits being considered. This aggregate genetic value (H) is given as:

$$H = a_i G_i$$

where a_i = economic value or relative value for trait i and G_i = genetic value for trait i.

For more details on selection index and other selection schemes to breed for multiple traits, readers must see Hallauer et al. (2010).

6.3. MOLECULAR BREEDING APPROACHES TO INTROGRESS MULTIPLE TARGET GENES GOVERNING SAME OR DIFFERENT TRAITS

Molecular breeding involves the use of molecular markers in combination with linkage maps and genomics to improve a particular trait. Various molecular breeding strategies include marker-assisted selection (MAS), marker-assisted backcrossing (MABC), marker-assisted recurrent selection (MARS), and genomic selection or genomic predictions (GS) (Ribaut et al., 2010). Marker-assisted selection involves indirect phenotypic selection in which individuals are selected based on the marker pattern. Readers can find more details on MAS in the following review papers, including those by Collard et al. (2005), Collard and Mackill (2008), Jena and Mackill (2008), Xu and Crouch (2008), and Gupta et al. (2010).

The general procedure of MAS is given in Fig. 6.3. Marker-assisted selection involves the following major methods: (1) screening of populations (e.g., F_2, F_3, recombinant inbred lines, double haploids, etc.) for genotypes of interest based on molecular markers, (2) marker-assisted backcross, where one or more genes/QTLs of interest are transferred from a donor parent to a recipient parent by repeated backcrossing to improve the target trait, (3) gene pyramiding schemes, where genes (two or more) identified in multiple lines/parents are accumulated into a single genotype, (4) marker-based recurrent selection, a complex scheme used for more loci involving several generations of selection and random mating of selected individuals, (5) selection based on an index combining molecular and phenotypic data, and (6) genomic selection, in which genomic estimated breeding value is obtained using information from genome-wide markers.

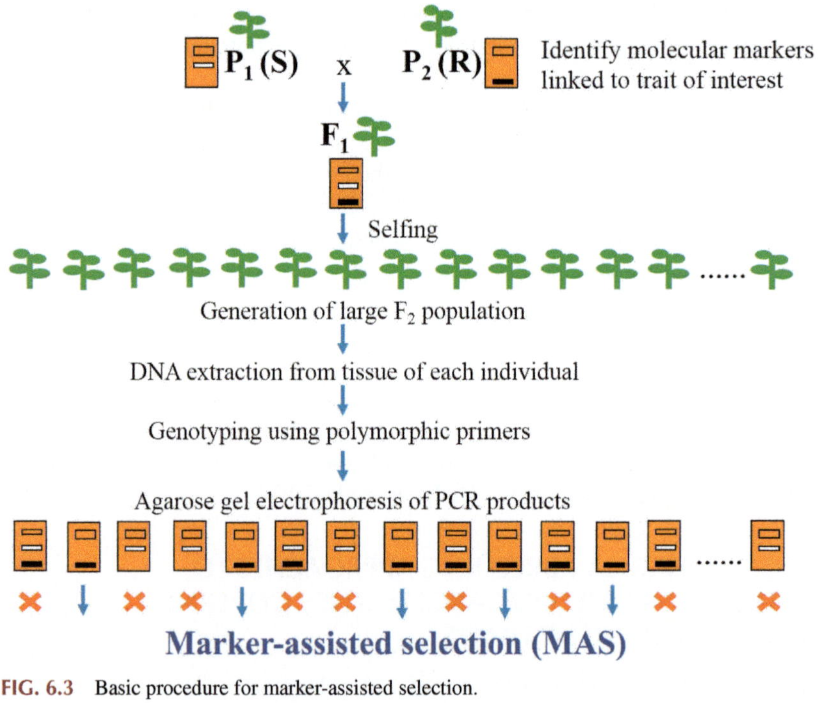

FIG. 6.3 Basic procedure for marker-assisted selection.

6.3.1 Molecular Marker Identification, Linkage Map Development and QTL Mapping

For a molecular breeding program, the availability and easy accessibility of genomic resources is a prerequisite. Technological advances have provided a range of resources like molecular markers, genetic linkage maps, whole genome sequences, transcriptomes, etc. The very first types of markers reported and used in lentils were morphological and isozyme markers (Zamir and Ladizinsky, 1984; Tadmor et al., 1987; Muehlbauer et al., 1989; Vaillancourt and Slinkard, 1993). Afterward, the information of different bases in the DNA molecule like point mutations and indels (insertion, deletion), a mutation in the repeat sequences were utilized to develop DNA-based markers in lentil. DNA markers have the advantages of abundance and high polymorphism over morphological and isozyme markers (Paterson et al., 1991), which can be generated to saturate a linkage map. Using the fragment length variation, restriction fragment length polymorphism (RFLP) markers were developed and were the first to be used for the construction of linkage map in lentils (Havey and Muehlbauer, 1989). Subsequently, PCR-based DNA markers, such as randomly amplified polymorphic DNA (RAPD) markers, were developed and used for diversity analysis, phylogenetic analysis, and for the identification of a taxonomic relationship among the members of genus *Lens* (Sharma et al., 1996; Ford et al., 1997; Ferguson et al., 2000), for linkage map construction (Eujayl et al., 1997, 1998a; Rubeena et al., 2003), for gene tagging (Eujayl et al., 1998b, 1999; Ford et al., 1999; Tullu et al., 2003), and for determining pathogen population structure (Ford et al., 2000). Amplified fragment length polymorphism (AFLP) markers have also been utilized for construction of genetic maps (Eujayl et al., 1998a; Durán et al., 2004; Hamwieh et al., 2005; Kahraman et al., 2004), to analyze genetic diversity in lentils (Sharma et al., 1996; Závodná et al., 2000), and to identify markers linked to traits (Tullu et al., 2003).

With the advantage of SSRs over other markers, these markers have been developed in lentils also. However, only two reports had been published that were related to SSR development in lentils (Závodná et al., 2000; Hamwieh et al., 2005, 2009), and some of them have been utilized for map construction (Durán et al., 2004; Hamwieh et al., 2005). More recently, a set of 122 and 360 new genomic SSR markers were reported by Verma et al., 2014; Andeden et al., 2015, respectively. Further, a large repertoire of 501 *Lens* SSR was developed by Verma et al., 2015 using two microsatellite genomic libraries enriched for (GA/CT) and (GAA/CTT) motif. Besides the above-reported markers, another class of markers has also been developed in lentils, intersimple sequence repeat (ISSR) markers, which are amplified with SSR-anchored primers and resistance gene analogue (RGA) markers. This type of marker is designed using the conserved region of the resistance genes of plants and has been used in mapping (Durán et al., 2004; Rubeena et al., 2003).

These PCR-based markers are being rapidly replaced by SNPs. Recently, a more advanced marker technology has been employed in lentils by Sharpe et al. (2013) in nine *L. culinaris* and two *L. ervoides* accessions using 454 pyrosequencing technology, identifying 1536 SNPs from a total of 44,879 SNPs using allele-specific illumina golden gate array and using them to construct an SNP-based genetic map of *L. culinaris* mapping population. Similarly, Temel et al. (2014) have identified another set of 50,960 SNPs and constructed a SNP-based linkage map in lentils. Since SNP discovery and genotyping require expensive and sophisticated platforms, the development and exploitation of SNP markers is still limited in lentils. More recently, Wong et al. (2015) developed an automated GBS pipeline and detected a total of 266,356 genome-wide SNPs for construction of a maximum-likelihood tree using 5389 SNPs for classification and characterization of species within the genus *Lens*. Readers can find more details on current knowledge in lentil genomics and its application for crop improvement in the excellent review paper by Kumar et al. (2015).

6.3.2 Genetic Linkage Maps of *Lens*

Genetic linkage maps have recently become cornerstones in the basic genetic analysis as well as in applied plant breeding. Linkage maps have assisted in the identification of DNA markers linked to major genes of agronomic importance and have permitted identification of tightly linked DNA tags for use as diagnostic tools in plant breeding. Through the use of linkage maps, characterization of quantitatively inherited traits has been facilitated, including identifying the genomic regions containing contributing loci, postulating the types of gene action that may be involved and determining the role of epistatic effects in specifying phenotype (Tanksley, 1993). Linkage maps based on molecular markers also have the potential to bridge the gap between the understanding of phenotype based on genetics and of organismal biochemistry and physiology (Gilpin et al., 1997). A detailed linkage map is required to define and distinguish QTL. Once major QTLs have been uncovered, tightly linked markers may be validated for use in marker-assisted selection (MAS) and potentially even as a starting point for the positional cloning of the underlying functional resistance gene(s) (Haley and Andersson, 1997).

The list of published maps including the most advanced map of lentil has been depicted in Table 6.3. The very first linkage map of lentil using DNA-based markers (RFLP) was constructed by Havey and Muehlbauer (1989). Today lentil linkage mapping has progressed dramatically with the advancement of molecular techniques.

6.3.3 QTL Mapping

The identification and localization of genes controlling variation in quantitative traits can greatly facilitate their selection in breeding programs. Thoday (1961) demonstrated that gene markers for simply inherited traits can be used as tags to

TABLE 6.3 List of Linkage Map Constructed in Lentils Using DNA-Based Molecular Markers

Cross Type	Mapping Population	Markers Mapped	No of Linkage Groups	Distance (cm)	Average Marker Density (cm)	Reference
Interspecific	F_2	20 RFLP, 14 others	9	333.0	9.79	Havey and Muehlbauer (1989)
Interspecific	F_2	28 RAPD, 1 RFLP, 4 others	9	206.1	6.24	Eujayl et al. (1997)
Interspecific	RIL	89 RAPD, 79 AFLP, 6 RFLP, 3 others	7	1073.0	6.0	Eujayl et al. (1998a)
Interspecific	F_2	64 DNA-based + others	10	560.0	8.75	Weeden et al. (1992)
Interspecific	F_2	38 RFLP, 38 others	10	–	–	Tahir et al. (1993)
Intraspecific	F_2	100 RAPD, 11 ISSR, 3 RGA	9	784.1	6.9	Rubeena et al. (2003)
Intraspecific (intersubspecific)	F_2	62 RAPD, 29 ISSR, 65 AFLP, 1 SSR, 4 others	10	2172.4	15.87	Durán et al. (2004)
Intraspecific	RIL	Total 130 (AFLP, RAPD, ISSR)	9	1192.0	9.1	Kahraman et al. (2004)
Intraspecific	RIL	49 AFLP, 39 SSR, 194 (110 RAPD, 80 AFLP, 4 others)	14	751.0	2.6	Hamwieh et al. (2005)
Intraspecific	F_5	97 ITAP, 18 SSR	7	928.4	14.74	Phan et al. (2007)
Intraspecific	RIL	207 markers (144 AFLP, 54 RAPD, and 9 SSRs)	12	1868	8.9	Tullu et al. (2008)
Intraspecific	RIL	Total 166 (RAPD, ISSR)	11	1396.3	8.4	Tanyolac et al. (2010)

Continued

TABLE 6.3 List of Linkage Map Constructed in Lentils Using DNA-Based Molecular Markers—cont'd

Cross Type	Mapping Population	Markers Mapped	No of Linkage Groups	Distance (cm)	Average Marker Density (cm)	Reference
Intraspecific	RIL	21 SSR, 27 RAPD, 89 SRAP	14	1565.2	11.3	Saha et al. (2010a)
Intraspecific	F_5	196 SSRs	11	1156.4	5.9	Gupta et al. (2011)
Intraspecific (intersubspecific)	F_2	28 SSR, 9 ISSR, 162 RAPD	11	3843.4	19.3	Gupta et al. (2012)
Intraspecific	RIL	563 SNP, 10 SSRs, and 4 others	7	697	1.2	Fedoruk et al. (2013)
Intraspecific	RIL	57 SSR, 261 SNP	10	1178.0	3.7	Kaur et al. (2013)
Intraspecific	RIL	6 SSR, 537 SNP	7	834.7	–	Sharpe et al. (2013)
Interspecific	F_2	377 gene-based markers	7	973.7	2.6	Verma et al. (2014)
Intraspecific	RIL	216 SSR	7	1183.7	5.4	Verma et al. (2015)
Intraspecific	RIL	4 SSRs and 1780 SNPs	7	4060.6	2.3	Ates et al. (2016)
Intraspecific	RIL	689 SNPs and SSRs	7	2429.6	3.5	Sudheesh et al. (2016)

locate quantitative trait loci (QTL). The technique for identification of QTL by gene markers became more efficient with the availability of molecular markers.

Molecular markers have been used to identify QTLs in several crops such as maize (Stuber and Moll, 1972; Stuber et al., 1982; Edwards et al., 1987; Liu et al., 2012; Almeida et al., 2012), wheat (Cuthbert et al., 2008; Rebetzke et al., 2008; Bennett et al., 2012), rice (Wei et al., 2012; Vikram et al., 2011; Steele et al., 2012; Dixit et al., 2012), tomatoes (Tanksley et al., 1982; Paterson et al., 1988), common beans (Blair et al., 2006; Kwak et al., 2008; Pérez-Vega et al., 2010), soybeans (Liu et al., 2005; Du et al., 2009; Abdel-Haleem et al., 2011; Zhang et al., 2012), and chickpeas (Anbessa et al., 2006; Lichtenzveig et al., 2006; Anbessa et al., 2009; Gowda et al., 2011).

Few QTL studies have been reported so far in lentils, as information on the genetic control of important quantitatively inherited traits in lentils such as plant height (PH) and days to flower (DTF) is limited. According to Tahir and Muehlbauer (1994), four QTLs for DTF were detected on LG1, LG2, LG4, and LG7, whereas four QTLs for PH were detected on LG1, LG2, LG3, and LG5 of the interspecific map of Muehlbauer et al. (1995). Later, Sarker et al. (1999) identified a recessive allele and a polygenic system to control days to flowering in lentil, and the flowering locus was assigned to LG5 of the interspecific map reported by Muehlbauer et al. (1995). Earliness is an adaptive trait and is one of the major factors of agronomic variation (Worland, 1996). Development of early maturing lines with optimum vegetative and reproductive phases combined with high and stable yield is a major goal in lentil breeding.

QTL studies using linkage mapping are abundant in nearly all crop species, including lentils. Multiple QTLs in lentils using both inter- and intraspecific maps have been identified and mapped, many of them being agronomically important traits such as plant height, days to flowering, winter hardiness, pod dehiscence, growth habit, and yield, and QTLs for resistance to diseases like ascochyta blight, anthracnose, and stemphylium blight (Ford et al., 1999; Rubeena et al., 2006; Tullu et al., 2008; Saha et al., 2010a; Phan et al., 2007). Table 6.4 shows agronomically important traits mapped in lentils. Even with the number of QTLs that have already been mapped, very few markers have been progressed to the MAS level in lentil breeding.

6.3.4 Marker-Assisted Backcrossing

Marker-assisted backcrossing is the simplest form of MAS. A backcross program aided with molecular markers is known as marker-assisted backcrossing (MABC). It aims at introgression of one or few target genes/QTLs of interest from donor line (may be agronomically inferior but contains desired genes for certain traits) into a desired genetic background (agronomically superior). The recipient parent act as a recurrent parent and the objective is to reduce the donor genome content in subsequent generations by repeated backcrossing to the recurrent parent. The MABC aims to (1) transfer the target trait from a

TABLE 6.4 Molecular Markers Closely Associated With Desirable Lentil Breeding Traits for Use in Marker-Assisted Selection

Trait Mapped	QTL/Gene	Associated Markers	Reference
Days to flower	QTL (4)	–	Tahir and Muehlbauer (1994)
Plant height	QTL (PH) (4)	–	Muehlbauer et al. (1995)
Fusarium wilt resistance	*Fw*	OPK15	Eujayl et al. (1998b)
Frost tolerance	*Frt*	OPS-16	Eujayl et al. (1999)
Ascochyta blight resistance	*AbR*1	RV01, RB18, SCARW19	Ford et al. (1999)
Ascochyta blight resistance	*ral2*	UBC227, OPD-10	Chowdhury et al. (2001)
Ascochyta blight resistance (mapped as a QTL)	QTL1, QTL2	C-TTA/M-AC	Rubeena et al. (2003)
	QTL3	M20	
Anthracnose resistance	*Lct2*	OPE06, UBC704	Tullu et al. (2003)
Cotyledon color	*Yc*	–	Durán et al. (2004)
Winter hardiness	–	UBC808-12	Kahraman et al. (2004)
Fusarium wilt resistance	*Fw*	SSR59-2B, p17m30710	Hamwieh et al. (2005)
Ascochyta blight resistance	–	ctcaccB, *LCt2*	Tullu et al. (2006)

No. of branches at the 1st node	BN	I890_2, M5D185; M7D235; I808_2, M6D134; M5D162, M7F231; OPC6_4, M6B121	Fratini et al. (2007)
Height of the 1st node	HN	M6F80, OPG10_3; M6D134, A15_2.1	
Total no. of branches	TB	I855_4, M6D179; M5D175, M6F62	
Plant height	PH	M8B234, OPA10_2; MGD131, OPH2_1; ms21, MS56;	
Flowering time	FT	I64_4.3, OPG10_4; M5D175, M6F62	
Dehiscence	DH	OPG3_1, WS_2.1; M7F63, OPH2_1	
Seed weight	SW	I864_5, I835_1; M7B264, OPP13_2; OPP16_1, M8B268	
Seed diameter	SD	OPW1_2, OPW16_3; I855_4, OPW19_3; M5D162, M7F231	
Plant height	QTL (PH)	SSR113, cacaggF	Tullu et al. (2008)
Earliness	QTL	SSR302, UBC 213b	
Leaf area	Leaf area	–	Kahraman et al. (2010)
Winter hardiness	WH-3	–	
Rust resistance gene	–	F7XEM4a	Saha et al. (2010a)
Stemphylium blight	$QLG4_{80–81}$	ME5XR10, ME4XR16c, and UBC34	Saha et al. (2010b)
Spotting	spotting	MCTAEACT_2, MCATEAAG_7;	Tanyolac et al. (2010)
Cotyledon color	cotcolor	Q10a, B10a	

Continued

TABLE 6.4 Molecular Markers Closely Associated With Desirable Lentil Breeding Traits for Use in Marker-Assisted Selection—cont'd

Trait Mapped	QTL/Gene	Associated Markers	Reference
Ascochyta blight	QTL1	DK 225-UBC825c	Gupta et al. (2012)
	QTL2	AC097a-V20a	
	QTL3	UBC890-ARG10	
	QTL4	ILMs25-UBC857b	
	QTL5	UBC855a-UBC830b	
	QTL6	UBC807a-Lup91	
Cotyledon color	*Yc*	LcC13114p356	Fedoruk et al. (2013)
Seed diameter	–	LcC02348p98, LcC04409p171, LcC05284p449, LcC05332p332, LcC05579p160	
Days to flowering	–	LcC06044p758, LcC09496p566, LcC23363p108	
Days to flower (Earliness)	QLG (3)	UBC34, UBC1, GLLC556, ME5XR7b, F8XEM58b	Saha et al. (2013)
	QLG (1)	SSR204b	
Seed diameter	QLG (3)	UBC34, UBC1	
100 seed weight	QLG (5)	UBC34, UBC1, UBC38b, UBC24a	
Boron tolerance	q_boron_IM	SNP_20000246, SNP_20002998	Kaur et al. (2013)
Rust resistance	–	GLLC 527	Dikshit et al. (2016)

Seed size	qSS	LcSSR426-LcSSR487	Verma et al. (2015)
Seed weight	qSW	LcSSR426-LcSSR280	
Selenium tolerance	SeQTL2.1	SNPT1002, SNPT2035	Ates et al. (2016)
	SeQTL5.2	SNPT2159, SNPT2359	
	SeQTL5.3	SNPT2312, SNPT1988	
	SeQTL5.1	SNPT756, SNPT1054, SNPT2242	
Seedling survival drought tolerance gene	*sdt*	PLC_105, PBA_LC_1480	Singh et al. (2016)
Ascochyta blight	AB_IH1	PBA_LC_0629–SNP_20005010	Sudheesh et al. (2016)
	AB_IH1.2	SNP_20002370–SNP_20002371	
	AB_NF1	SNP_20001370–SNP_20001765	
	AB_IH1	SNP_20005010–SNP_20004695	
	AB_IH1.3	SNP_20000505–SNP_20000553	

donor parent into the recurrent parent based on markers flanked to target gene/QTL (foreground selection), and (2) recover the recipient parent genome to its maximum (background selection) based on markers distributed throughout the genome. The efficiency of MABC depends on the genetic distance of markers from the target gene, a number of markers used for a target gene, the population size of each backcross generation, recurrent parent background, and undesirable linkages. The general procedure of MABC (Fig. 6.4), for example, for ascochyta blight resistance gene transfer, can be followed in lentils as given below.

The procedure, in general, involves (i) selection of markers well distributed across the genome for background selection and tightly flanked markers on each side of target gene for foreground selection, (ii) crossing between donor (resistant to ascochyta blight) and recipient parent (recurrent parent) to generate the F_1, which is then backcrossed to recurrent parent, (iii) genotyping of BC_1F_1 plants based on foreground and background markers to select for target gene and recovery of background genome, (iii) repeating steps (ii) and (iii) until to produce enough BC_3F_1 seeds (200–500), and (iv) selfing and genotyping of all the selfed progenies homozygous for resistance gene. Bulk all the homozygous, resistant plants for seed increase and trait evaluation.

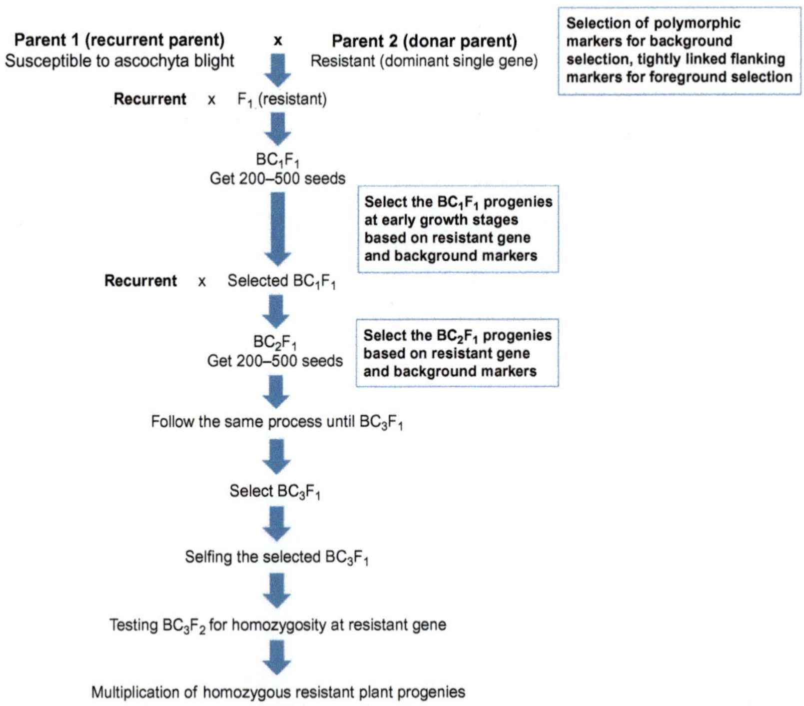

FIG. 6.4 General schematic representation of marker-assisted backcrossing for a single gene. For two or three genes, the same process can be followed, but large population size is required at each backcross generation.

Marker-assisted backcrossing is advantageous when phenotyping of a trait is difficult or expensive, the heritability of target trait is low, the trait is expressed in late stages of plant development, or traits controlled by a recessive gene or multiple genes need to combine for one or more traits.

6.3.5 Gene Pyramiding or Gene Stacking

Breeders are often interested in transferring or introgressing many genes from different sources into a desirable variety for genetic improvement. In lentils, when developing elite lines, breeders need to combine traits from multiple parents, particularly for resistance to biotic and abiotic stresses. The process of combining traits is known as gene pyramiding, and the concept was proposed by Nelson (1978) to develop crop varieties with few to several different oligo genes for durable disease resistance. Gene pyramiding is a method in which many desirable genes from different parents are assembled into a single genotype. Gene pyramiding has also been called as multitrait introgression, as often, genes governing two or more traits are introgressed into a single recurrent parent. The introgression of multiple QTLs/genes and its effects have been proposed in crop species like wheat, barley, rice, and soybean (Richardson et al., 2006; Jiang et al., 2007a,b; Wang et al., 2012; Luo et al., 2016). However, in lentils, there is only one report of gene pyramiding, by Tarán et al. (2003). They used molecular markers to pyramid genes for resistance to ascochyta blight and anthracnose in lentil.

Pyramiding of multiple QTLs/genes may be conducted through the multiple-parent crossing, backcrossing, and recurrent section. Gene pyramiding aided with molecular markers depends upon the number of genes/QTLs, the number of parents containing the target genes/QTLs, the heritability of target genes/QTLs, marker-target gene associations, duration needed to complete the gene assembly, and relative cost. Pyramiding three or four genes can be achieved through three-way, four-way, or double crossing. However, if more than four genes/QTLs are to be pyramided, complex or multiple crossing or recurrent selection schemes may often be preferred.

6.3.6 Gene Pyramiding Through Multiple Parents Crossing

In this breeding strategy, the goal is to accumulate genes/QTLs identified in multiple parents into a single genotype. Use of molecular markers aids in complete gene/QTL identification of the progeny at each generation, thus, enhancing the speed of the pyramiding process. The gene pyramiding scheme can be divided into two parts. The first part aims to accumulate one copy of all target genes in a single genotype called the root genotype. The second part, called the fixation step, aims to fix all the target genes into a homozygous state to derive the ideal genotype. Servin et al. (2004) called these two parts the pedigree and fixation steps, respectively. An example of a gene pyramiding scheme for accumulating six target genes derived from six parental lines each containing

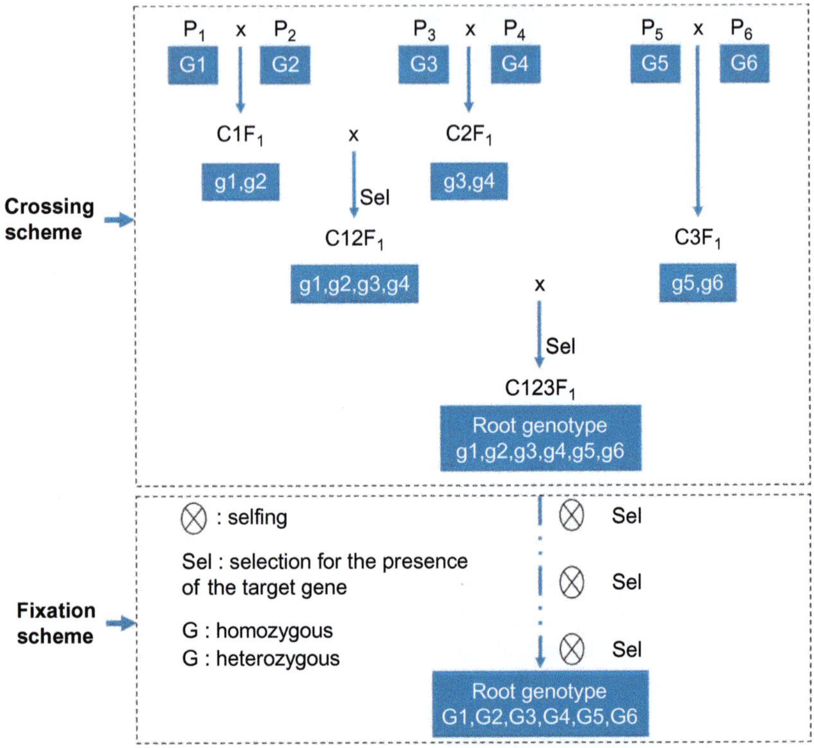

FIG. 6.5 Example of a gene-pyramiding scheme accumulating six target genes.

the target gene is depicted in Fig. 6.5. For more details on gene pyramiding, readers should see the reviews by Servin et al. (2004), Joshi and Nayak (2010), and Ye and Smith (2008).

6.3.7 Marker-Assisted Backcrossing-Gene Pyramiding

For marker-assisted backcross-based gene pyramiding, three breeding schemes can be used: (1) stepwise transfer (Fig. 6.6), (2) simultaneous transfer (Fig. 6.7), and (3) convergent backcrossing or transfer (Fig. 6.8). In stepwise backcrossing, target genes are transferred into the recurrent parent in order (Fig. 6.6). The advantage of this method is that it is more precise and easier to implement as only one gene/QTL is involved at one time, and thus, the population size and genotyping amount will be small. The disadvantage is that it takes a longer time to complete the pyramiding process. The advantage of simultaneous or synchronized backcrossing is that it takes less time to complete (Fig. 6.7). However, in backcrossing, all target genes/QTLs are involved at the same time, and thus, it requires a large population and more genotyping. Convergent backcrossing is a strategy combining the advantages of stepwise and synchronized backcrossing

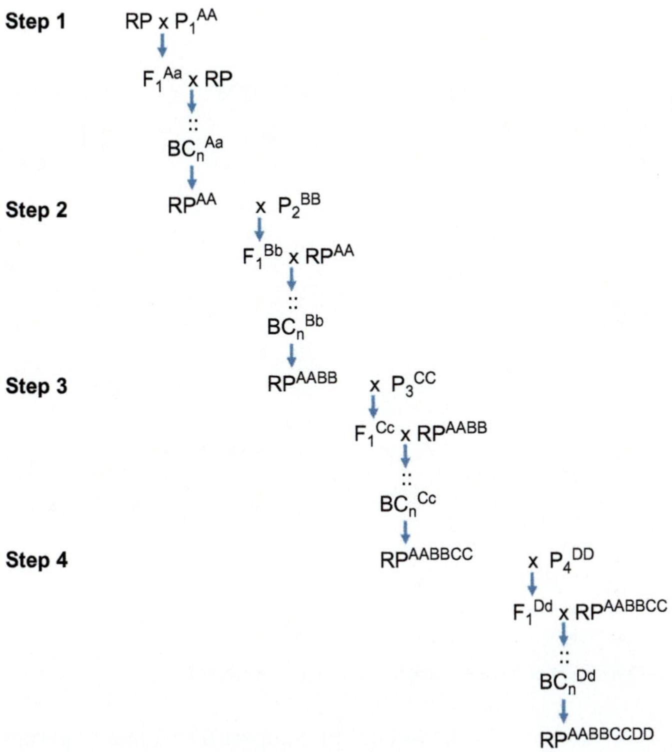

FIG. 6.6 Stepwise backcrossing to pyramid multiple genes.

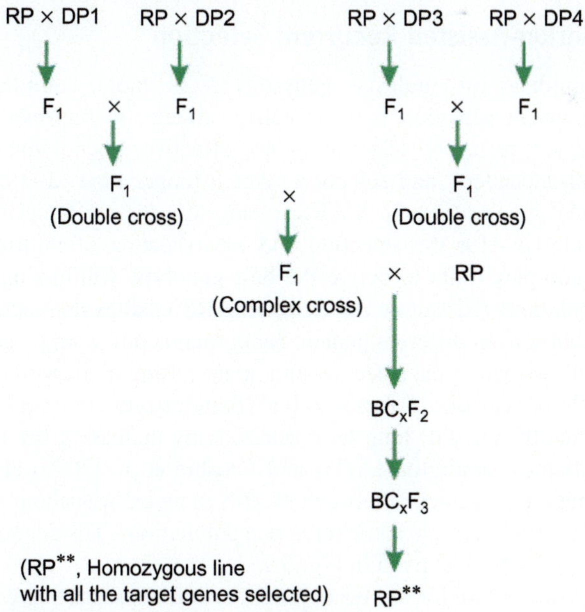

FIG. 6.7 Simultaneous backcrossing for multiple genes.

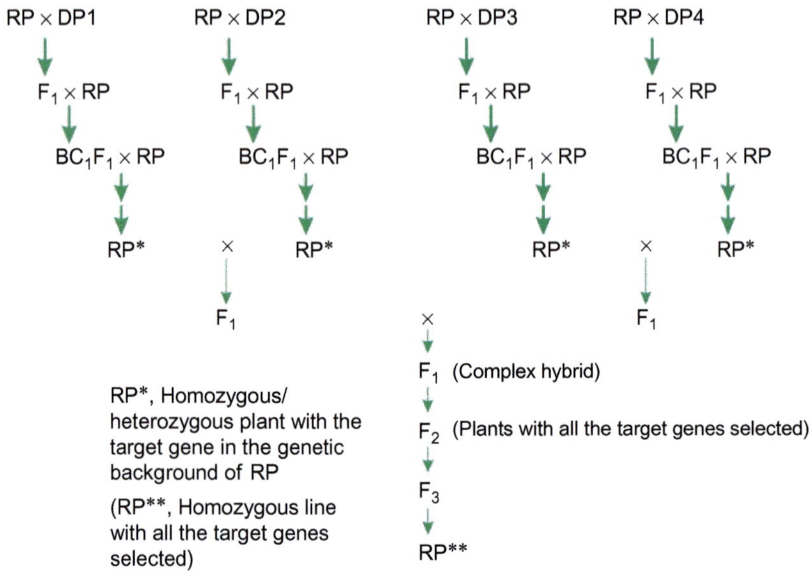

FIG. 6.8 Convergent or separate backcrossing of multiple genes.

(Fig. 6.8). Convergent backcrossing, in comparison, is more appropriate and acceptable, as in this scheme, not only is time reduced, but gene fixation and pyramiding are also more easily assured.

6.3.8 Marker-Assisted Recurrent Selection

Gene pyramiding for multiple genes/QTLs is more complex and less proven. Recurrent selection is an effective strategy to improve polygenetic traits. However, recurrent selection is not effective, phenotypic selection is environment-dependent, and selection takes a longer time (2–3 crop seasons for one cycle of selection). In MARS, recurrent selection is performed using molecular markers for the selection and identification of multiple genomic regions of complex traits to derive the best genotype within single or across related populations (Ribaut et al., 2010). MARS enables the accumulation of favorable alleles from different genetic backgrounds into a single genetic background. It allows genotypic selection and intercrossing in the same crop season for one cycle of selection. Johnson (2004) demonstrated in maize that MARS increased the efficiency of long-term selection by increasing the frequency of favorable alleles. Eathington (2005) and Crosbie et al. (2006) also indicated that the genetic gain achieved through MARS in maize was about twice that of phenotypic selection (PS) in some reference populations. The general procedure for performing MARS is given in Fig. 6.9.

The basic procedure for MARS requires several steps, starting with selection of the parents from the same or different populations. Population development

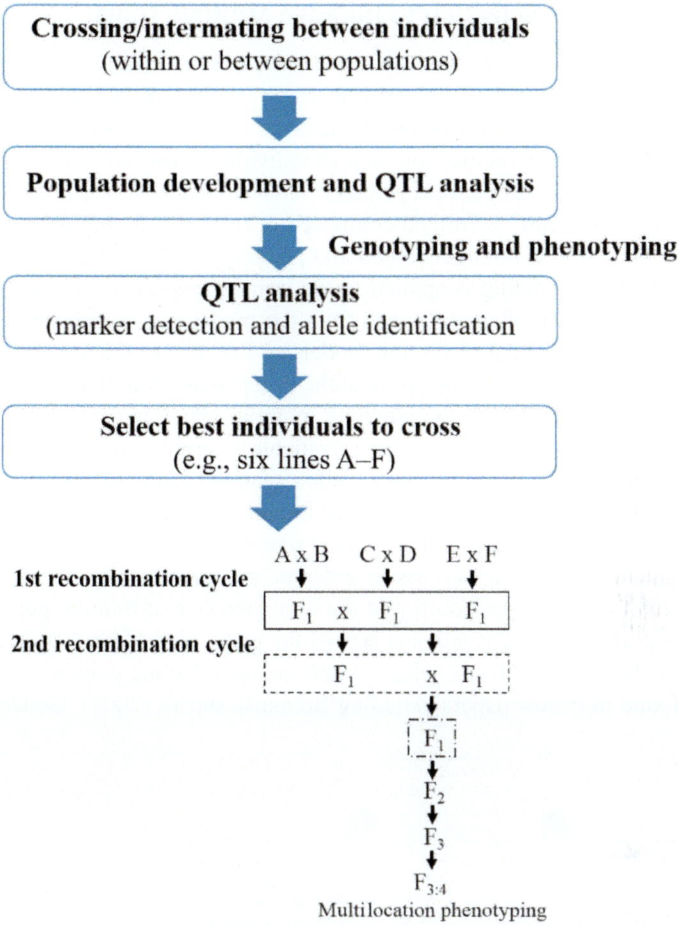

FIG. 6.9 The general schematic representation of marker-assisted recurrent selection (MARS) in crops.

for MARS F_3-derived populations is generally sufficient and can be advanced through a single-seed descent method for seed increase to perform multiplication testing. A population size of 200 to 500 is desired, which will also depend upon the precision of QTL mapping. QTL analysis can be performed after genotyping and phenotyping to detect the markers and favorable alleles. For final QTL selection, index selection (based on both phenotype and marker data) with different weight given to various key traits, can be helpful. Recombination breeding (or cycle) is performed once key QTLs are identified. The best individuals are selected after genotyping to use in the recombination cycle. An example of four lines crossed to create 2 pair progenies, with the one pair of resulting F_1 in a second recombination cycle. At each stage, genotyping is done to select the best F_{1s} and is used again for the next cycle of recombination.

6.3.9 Genomic Selection to Improve Multiple Traits

Genomic selection or genome-wide selection is a specific case of marker-assisted selection in which information from markers distributed across the whole genome is used in selection. It involves statistical modeling and novel bioinformatics tools to predict how well an individual plant is performing before testing in the field. To implement genomic selection in a breeding program, first a "training population" is formed composed of related breeding material. The training population is both genotyped and phenotyped for all traits of interest. Then, statistical modeling is applied to the training population sample to predict the performance of the lines and train the model. It is necessary to apply the appropriate statistical prediction models so that the accuracy of prediction is high, and future predictions are reliable. After model training, the best prediction model is applied to the breeding material which has been only genotyped with no phenotypic information available to predict the performance of lines and estimate the breeding value, called genomic estimated breeding value (GEBV). The lines with higher GEBV can be selected as the best lines for future breeding programs. The general schematic representation of genomic selection is given in Fig. 6.10. The success of genomic selection largely depends upon the nature of traits, the heritability of the traits, the size of training population, marker nature and density, genome size of the plant, type of prediction model, etc. More details on genomic selection and factors affecting genomic selection can be found in review papers including Bernardo and Yu (2007), Goddard and

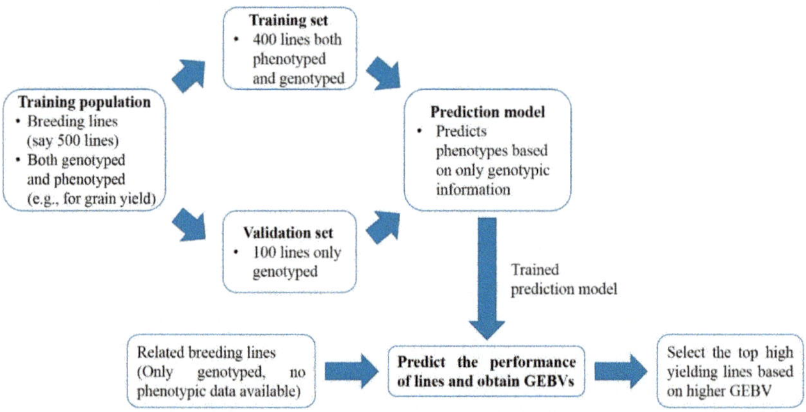

FIG. 6.10 A simple schematic representation of genomic selection procedure. First, a training population is created which has both phenotypic and genotypic information. The training population is then divided into a training set and validation set to apply various statistical prediction models and predict the performance of lines. The higher the prediction accuracy, the better the prediction model. After training the prediction model, the appropriate model is applied to breeding lines that have been only genotyped with no phenotypic data available to estimate the breeding values, called genomic estimated breeding values (GEBVs), using genotypic (marker) information only. The lines with higher GEBVs are then selected and may go in future breeding programs, or the best line can be selected to release as a variety.

Hayes (2007), Heffner et al. (2010), Lorenz et al. (2011), Xu et al. (2012), and Desta and Ortiz (2014).

To date, no information on genomic selection or genomic prediction is available in lentil crop. Several studies in legumes and cereals have shown great potential for genomic selection in line breeding to enhance the selection for major agronomic traits like yield (Jarquín et al., 2014; Tayeh et al., 2015; He et al., 2016; Michel et al., 2016). In the future, genomic selection has great potential and promises to revolutionize lentil breeding. The ability to select based on genomic predictions, rather than on phenotypic observations, will result in a rapid increase in genetic gains and efficiency in lentil breeding. This dramatic increase in genetic gains will allow lentil breeders to meet future food demands.

However, for successful lentil improvement, the focus should not be on a single trait, but on multiple traits. Plant breeders often recorded phenotypic data on multiple traits including yield and yield components, quality traits, and reaction to biotic and abiotic stresses. To improve multiple traits simultaneously, it is important that a favorable genetic correlation exists between the traits. Genetic correlation indicates the genetic information one trait carries for other traits. Genetic correlation between traits is the basis for multitrait genomic selection. The success of multitrait genomic selection largely depends upon the correlation between the traits and selection of appropriate statistical prediction models to handle multiple traits (Jia and Jannink, 2012). Multitrait genomic selection was originally designed to benefit from the information contained in correlated traits (Calus and Veerkamp, 2011). Since then, this approach has been applied in crops to simultaneously improve traits, but studies on this are limited (Schulthess et al., 2016).

As discussed earlier in the section, while improving several quantitative traits simultaneously, selection index (SI) is more efficient than independent culling or tandem selection, particularly when the traits are negatively correlated. The selection indices take into account the overall performance of genotypes and, thus, consider the ability of favorable levels for some trait(s) to compensate for unfavorable levels in another trait(s) (Dolan et al., 1996). In crops, the applicability of various selection indices has been evaluated to improve the overall genotype performance according to different traits and to select the best-performing genotype (Elgin et al., 1970; Eagles and Frey, 1974; Suwantaradon et al., 1975; Openshaw and Hadley, 1984; Holbrook et al., 1989; Dolan et al., 1996; Ceron-Rojas et al., 2015 and Cerón-Rojas et al., 2016). In the past, selection indices for multiple trait selections have been coupled with information on molecular markers controlling QTLs (Cerón-Rojas et al., 2016). Selection indices on multiple traits using information on genome-wide markers as in genomic selection have also been studied (Bernardo, 2010; Dekkers, 2007; Schulthess et al., 2016). However, such studies have not been extensively performed in plants. In lentils, there is a great opportunity to take advantage of genomic selection, which may be an effective and efficient novel tool to improve multiple traits simultaneously.

6.4. CONCLUSIONS

In lentil breeding, it is necessary to breed for multiple traits including yield and quality as well as tolerance to abiotic and biotic stresses in order to develop successful cultivars to meet consumers' demands under challenging farming conditions. Conventional breeding methodologies to integrate multiple traits into one genetic background have great potential in lentil breeding, though it is challenging. Breeders must prioritize target traits and availability of resources, cost, and time to breed for multiple traits. Molecular breeding technologies including marker-assisted selection, marker-assisted backcrossing, gene pyramiding, and genomic selection can greatly supplement conventional breeding and help in multiple traits or gene integration. Molecular breeding approaches present great opportunities and prospects in lentils, and all molecular breeding tools are still in their infancy in lentil crop. There is a constant need to develop effective genomic resources in lentils, including high-throughput marker identification and development, high-density linkage maps, and gene identification methodologies. Genomic selection, a future molecular breeding tool, has great potential in lentil breeding to increase the efficiency of selection and improve multiple traits simultaneously.

REFERENCES

Abbo, S., Ladizinsky, G., Weeden, N.F., 1991. Genetic analysis and linkage study of seed weight in lentil. Euphytica 58, 259–266.

Abdel-Haleem, H., Lee, G.J., Boerma, R.H., 2011. Identification of QTL for increased fibrous roots in soybean. Theor. Appl. Genet. 122, 935–946.

Aghili, P., Imani, A.A., Shahbazi, H., Alaei, Y., 2012. Study of correlation and relationships between seed yield and yield components in Lentil (Lens culinaris Medik.). Ann. Biol. Res. 3, 5042–5045.

Almeida, G.D., Makumbi, D., Magorokosho, C., Nair, S., Borém, A., Ribaut, J.M., Bänziger, M., Prasanna, B.M., Crossa, J., Babu, R., 2012. QTL mapping in three tropical maize populations reveals a set of constitutive and adaptive genomic regions for drought tolerance. Theor. Appl. Genet. https://doi.org/10.1007/s00122-012-2003-7.

Anbessa, Y., Warkentin, T., Vandenberg, A., Bandara, M., 2006. Heritability and predicted gain from selection in components of crop duration in divergent chickpea cross populations. Euphytica 152, 1–8.

Anbessa, Y., Taran, B., Warkentin, T.D., Tullu, A., Vandenberg, A., 2009. Genetic analyses and conservation of QTL for ascochyta blight resistance in chickpea (*Cicer arietinum* L.). Theor. Appl. Genet. 119, 757–765.

Andeden, E.E., Baloch, F.S., Cadir, E., Toklu, L., Ozkan, H., 2015. Development, characterization and mapping of microsatellite markers for lentil (*Lens culinaris* Medik.). Plant Breed. 134, 589–598.

Anjam, M.S., Ali, A., Iqbal, S.H.M., Haqqani, A.M., 2005. Evaluation and correlation of economically important traits in exotic germplasm of lentil. Int. J. Agric. Biotechnol. 7 (6), 959–961.

Ashrie, A.A.M., Fares, W.M., Selim, T., Fateh, S.A., 2012. The relationships among some yield characters of eight lentil genotypes. Egypt. J. Plant Breed. 16, 107–119.

Ates, D., Sever, T., Aldemir, S., Yagmur, B., Temel, H.Y., Kaya, H.B., Alsaleh, A., Kahraman, A., Ozkan, H., Vandenberg, A., Tanyolac, B., 2016. Identification QTLs controlling genes for se uptake in lentil seeds. PLoS ONE 11 (3), e0149210https://doi.org/10.1371/journal.pone.0149210.

Bakshs, A., Ghafoor, A., Malik, B.A., 1993. Genetic variability and correlation in lentil. Pak. J. Agric. Res. 14 (2/3), 246–250.

Bennett, D., Reynolds, M., Mullan, D., Izanloo, A., Kuchel, H., Langridge, P., Schnurbusch, T., 2012. Detection of two major grain yield QTL in bread wheat (*Triticum aestivum* L.) under heat, drought and high yield potential environments. Theor. Appl. Genet. 125, 1473–1485.

Bernardo, R., 2001. What if we knew all the genes for a quantitative trait in hybrid crops? Crop Sci. 41, 1–4.

Bernardo, R.N., 2010. Breeding for Quantitative Traits in Plants. Stemma Press, Woodbury.

Bernardo, R., Yu, J., 2007. Prospects for genome-wide selection for quantitative traits in maize. Crop Sci. 41, 1082–1090.

Blair, M.W., Iriarte, G., Beebe, S., 2006. QTL analysis of yield traits in an advanced backcross population derived from a cultivated Andean x wild common bean (*Phaseolus vulgaris* L.) cross. Theor. Appl. Genet. 112, 1149–1163.

Calus, M.P.L., Veerkamp, R.F., 2011. Accuracy of multi-trait genomic selection using different methods. Genet. Sel. Evol. 43, 26–32.

Ceron-Rojas, J.J., Crossa, J., Arief, V.N., Basford, K., Rutkoski, J., Jarquin, D., Alvarado, G., Beyene, Y., Semagn, K., DeLacy, I., 2015. A genomic selection index applied to simulated and real data. Genes Genom. Genet. 5 (10), 2155–2164.

Cerón-Rojas, J.J., Crossa, J., Toledo, F.H., Sahagún-Castellanos, J., 2016. A predetermined proportional gains eigen selection index method. Crop Sci. 56, 2436–2447.

Chowdhury, M.A., Andrahennadi, C.P., Slinkard, A.E., Vandenberg, A., 2001. RAPD and SCAR markers for resistance to ascochyta blight in lentil. Euphytica 118, 331–337.

Collard, B.C.Y., Mackill, D.J., 2008. Marker-assisted selection: an approach for precision plant breeding in the twenty-first century. Philos. Trans. R. Soc. Lond. Ser. B Biol. Sci. 363, 557–572.

Collard, B.C.Y., Jahufer, M.Z.Z., Brouwer, J.B., et al., 2005. An introduction to markers, quantitative trait loci (QTL) mapping and marker-assisted selection for crop improvement: the basic concepts. Euphytica 142, 169–196.

Crosbie, T.M., Eathington, S.R., Johnson, G.R., Edwards, M., et al., 2006. Plant breeding: past, present and future. In: Lamkey, K.R., Lee, M. (Eds.), Plant Breeding: The Arnel R. Hallauer International Symposium. Blackwell Publishing, Oxford, UK, pp. 3–50.

Cubero, J.I., Perez de la Vega, M., Fratini, R., 2009. Origin, phylogeny, domestication and spread. In: Erskine, W., Muehlbauer, F.J., Sarker, A., Sharma, B. (Eds.), The Lentil: Botany, Production and Use. CABI, Cambridge, pp. 13–33.

Cuthbert, J.L., Somers, D.J., Brule-Babel, A.L., Brown, D., Crow, G.H., 2008. Molecular mapping of quantitative trait loci for yield and yield components in spring wheat (*Triticum aestivum* L.). Theor. Appl. Genet. 117, 595–608.

Dekkers, J.C.M., 2007. Prediction of response to marker-assisted and genomic selection using selection index theory. J. Anim. Breed. Genet. 124, 331–341.

Desta, Z.A., Ortiz, R., 2014. Genomic selection: genome-wide prediction in plant improvement. Plant Sci. 19 (9).

Dikshit, H.K., Singh, A., Singh, D., Aski, M., Jain, N., Hegde, V.S., Basandrai, A.K., Basandrai, D., Sharma, T.R., 2016. Tagging and mapping of SSR marker for rust resistance gene in lentil (*Lens culinaris* Medikus subsp culinaris). Indian J. Exp. Biol. 54, 394–399.

Dixit, S., Swamy, B.P.M., Vikram, P., Ahmed, H.U., Cruz, M.T.S., Amante, M., Atri, D., Leung, H., Kumar, A., 2012. Fine mapping of QTLs for rice grain yield under drought reveals sub-QTLs conferring a response to variable drought severities. Theor. Appl. Genet. 125, 155–169.

Dolan, D.J., Stuthman, D.D., Kolb, F.L., Hewings, A.D., 1996. Multiple trait selection in a recurrent selection population in oat (*Avena sativa* L.). Crop Sci. 36, 1207–1211.

Du, W., Yu, D., Fu, S., 2009. Detection of quantitative trait loci for yield and drought tolerance traits in soybean using a recombinant inbred line population. J. Integr. Plant Biol. 51, 868–878.

Dudley, J.W., 1993. Molecular markers in plant improvement: manipulation of genes affecting quantitative traits. Crop Sci. 33, 660–668.

Durán, Y., Fratini, R., Garcia, P., de la Vega, M.P., 2004. An intersubspecific genetic map of *Lens*. Theor. Appl. Genet. 108, 1265–1273.

Eagles, H.A., Frey, K.J., 1974. Expected and actual gains in economic value of oat lines from five selection methods. Crop Sci. 14, 861–864.

Eathington, S.R., 2005. In: Practical applications of molecular technology in the development of commercial maize hybrids. Proceedings of the 60th Annual Corn and Sorghum Seed Research Conferences. American Seed Trade Association, Washington, DC.

Edwards, M.D., Stuber, C.W., Wendel, J.F., 1987. Molecular-marker-facilitated investigations of quantitative-trait loci in maize. I. Numbers, genomic distribution and types of gene action. Genetics 116, 113–125.

Elgin, J.H., Hill, R.R., Zeiders, K.E., 1970. Comparison of four methods of multiple trait selection for five traits in alfalfa. Crop Sci. 10, 190–193.

Erskine, W., Sarker, A., 2004. Lentil. In: Corke, H., Walker, C.E. (Eds.), Encyclopedia of Grain Sciences. Elsevier, London, UK, pp. 142–150.

Eujayl, I., Baum, M., Erskine, W., Pehu, E., Muehlbauer, F.J., 1997. The use of RAPD markers for lentil genetic mapping and the evaluation of distorted F_2 segregation. Euphytica 96, 405–412.

Eujayl, I., Baum, M., Powell, W., Erskine, W., Pehu, E., 1998a. A genetic linkage map of lentil (*Lens* sp.) based on RAPD and AFLP markers using recombinant inbred lines. Theor. Appl. Genet. 97, 83–89.

Eujayl, I., Erskine, W., Bayaa, B., Baum, M., Pehu, E., 1998b. Fusarium vascular wilt in lentil: inheritance and identification of DNA markers for resistance. Plant Breed. 117, 497–499.

Eujayl, I., Erskine, W., Baum, M., Pehu, E., 1999. Inheritance and linkage analysis of frost injury in lentil. Crop Sci. 39, 639–642.

FAOSTAT, 2016. Food and Agriculture Organization of the United Nations (FAO) Statistical Databases. http://faostat.fao.org/.

Fedoruk, M.J., Vanderberg, A., Bett, K.E., 2013. Quantitative trait loci analysis of seed quality characteristics in lentil using single nucleotide polymorphism markers. Plant Genome 6 (3), 37–39.

Ferguson, M.E., Maxted, N., van Slageren, M., Robertson, L.D., 2000. A re-assessment of the taxonomy of *Lens* Mill. (Leguminosae, Papilionoideae, Vicieae). Bot. J. Linn. Soc. 133, 41–59.

Fikru, M., Mekbib, F., Kumar, S., Ahmed, S., Chahota, R.K., Sharma, T.R., Singh, S., Gill, R.K., Kumar, A., 2014. Identification of molecular markers associated with rust (*Uromyces viciafabae* Pers.) resistance genes in lentil (*Lens culinaris* sub sp. *culinaris*). Can. J. Plant Prot. 2 (2), 27–36.

Ford, R., Pang, E.C.K., PWJ, T., 1997. Diversity analysis and species identification in *Lens* using PCR-generated markers. Euphytica 96, 247–255.

Ford, R., Pang, E.C.K., Taylor, P.W.J., 1999. Genetics of resistance to ascochyta blight (*Ascochyta lentis*) of lentil and identification of closely linked RAPD markers. Theor. Appl. Genet. 98, 93–98.

Ford, R., Garnier-Géré, P., Nasir, M., Taylor, P.W.J., 2000. The structure of *A. lentis* in Australia revealed with RAPD markers. Aust. Plant Pathol. 29, 36–45.

Fratini, R., Durán, Y., García, P., Pérez de la, V.M., 2007. Identification of quantitative trait loci (QTL) for plant structure, growth habit and yield in lentil. Span. J. Agric. Res. 5, 348–356.

Frederick, M., Cho, S., Sarker, A., McPhee, K., Coyne, C., Rajesh, P., Ford, P., 2006. Application of biotechnology in breeding lentil for resistance to biotic and abiotic stress. Euphytica 147, 149–165.

Gilpin, B.J., McCallum, J.A., Frew, T.J., Timmerman-Vaughn, G.M., 1997. A linkage map of the pea (*Pisum sativum* L.) genome containing cloned sequences of known function and expressed sequence tags (ESTs). Theor. Appl. Genet. 95, 1289–1299.

Goddard, M.E., Hayes, B.J., 2007. Genomic selection. J. Anim. Breed. Genet. 124, 323–330.

Gowda, S.J., Radhika, P., Mhase, L.B., Jamadagni, B.M., Gupta, V.S., Kadoo, N.Y., 2011. Mapping of QTLs governing agronomic and yield traits in chickpea. J. Appl. Genet. 52, 9–21.

Gupta, P.K., Balyan, H.S., Sharma, P.C., Ramesh, B., 1996. Microsatellites in plants: a new class of molecular markers. Curr. Sci. 70, 45–54.

Gupta, P.K., Kumar, J., Mir, R.R., 2010. Marker-assisted selection as a component of conventional plant breeding. Plant Breed. Rev. 33, 145–217.

Gupta, D., Taylor, P.W.J., Inder, P., Phan, H.T.T., Ellwood, S.R., Mathur, P.N., Sarker, A., Ford, R., 2011. Integration of EST-SSR markers of *Medicago truncatula* into intraspecific linkage map of lentil and identification of QTL conferring resistance to ascochyta blight at seedling and pod stages. Mol. Breed. 30, 429–439.

Gupta, M., Verma, B., Kumar, N., Chahota, R.K., Rathour, R., Sharma, S.K., Bhatia, S., Sharma, T.R., 2012. Construction of intersubspecific molecular genetic map of lentil based on ISSR, RAPD and SSR markers. J. Genet. 91, 279–287.

Haley, C.A., Andersson, L., 1997. Linkage mapping of quantitative trait loci in plants and animals. In: Dear, P.H. (Ed.), Genome Mapping—A Practical Approach. Oxford University Press, New York, pp. 49–71.

Hallauer, A.R., Carena, M.J., Filho, J.B., 2010. Quantitative Genetics in Maize Breeding. Springer, New York/Dordrecht Heidelberg London.

Hamwieh, A., Udupa, S.M., Choumane, W., Sarker, A., Dreyer, F., Jung, C., Baum, M., 2005. A genetic linkage map of *Lens* sp. based on microsatellite and AFLP markers and the localization of Fusarium vascular wilt resistance. Theor. Appl. Genet. 110, 669–677.

Hamwieh, A., Udupa, S.M., Sarker, A., Jung, C., Baum, M., 2009. Development of new microsatellite markers and their application in the analysis of genetic diversity in lentils. Breed. Sci. 59, 77–86.

Havey, M.J., Muehlbauer, F.J., 1989. Linkages between restriction fragment length, isozyme, and morphological markers in lentil. Theor. Appl. Genet. 77, 395–401.

He, S., Schulthess, A.W., Mirdita, V., et al., 2016. Genomic selection in a commercial winter wheat population. Theor. Appl. Genet. 129, 641–651.

Heffner, E.L., et al., 2010. Plant breeding with genomic selection: gain per unit time and cost. Crop Sci. 50, 1681–1690.

Holbrook, C.C., Burton, J.W., Carter, T.E., 1989. Evaluation of recurrent restricted index selection for increasing yield while holding seed protein constant in soybean. Crop Sci. 29, 324–329.

Jarquín, D., Crossa, J., Lacaze, X., et al., 2014. A reaction norm model for genomic selection using high-dimensional genomic and environmental data. Theor. Appl. Genet. 127, 595–607.

Jena, K.K., Mackill, D.J., 2008. Molecular markers and their use in marker-assisted selection in rice. Crop Sci. 48, 1266–1276.

Jia, Y., Jannink, J.L., 2012. Multiple-trait genomic selection methods increase genetic value prediction accuracy. Genetics 192, 1513–1522.

Jiang, G.L., Shi, J., Ward, R.W., 2007a. QTL analysis of resistance to Fusarium head blight in the novel wheat germplasm CJ 9306. I. Resistance to fungal spread. Theor. Appl. Genet. 116, 3–13.

Jiang, G.L., Dong, Y., Shi, J., Ward, R.W., 2007b. QTL analysis of resistance to *Fusarium* head blight in the novel wheat germplasm CJ 9306. II. Resistance to deoxynivalenol accumulation and grain yield loss. Theor. Appl. Genet. 115, 1043–1052.

Johnson, R., 2004. Marker assisted selection. In: Jannick, J. (Ed.), Plant Breeding Review. vol. 24(1), pp. 293–309.

Joshi, R.K., Nayak, S., 2010. Gene pyramiding—a broad spectrum technique for developing durable stress resistance in crops. Biotechnol. Mol. Biol. Rev. 5 (3), 51–60.

Kahraman, A., Kusmenoglu, I., Aydin, N., Aydogan, A., Erskine, W., Muehlbauer, F.J., 2004. QTL mapping of winter hardiness genes in lentil. Crop Sci. 44, 13–22.

Kahraman, A., Demirel, U., Ozden, M., Muehlbauer, F.J., 2010. Mapping of QTLs for leaf area and the association with winter hardiness in fall-sown lentil. Afr. J. Biotechnol. 9 (50), 8515–8519.

Karadavut, U., Kavurmaci, Z., 2013. Phenotypic and genotypic correlation for some characters in Lentil (*Lens culinaris* Medik.). Res. J. Agric. Environ. Manag. 2, 365–370.

Kaur, S., Cogan, N.O., Stephens, A., Noy, D., Butsch, M., Forster, J.W., Materne, M., 2013. EST-SNP discovery and dense genetic mapping in lentil (*Lens culinaris* Medik.) enable candidate gene selection for boron tolerance. Theor. Appl. Genet. 127, 703–713.

Kay, D., 1979. Food legumes. In: Tropical Development and Research Institute (TPI). TPI crop and product digest no. 3, UK, pp. 48–71.

Kumar, S., Rajendran, K., Kumar, J., Hamweih, A., Baum, M., 2015. Current knowledge in lentil genomics and its application for crop improvement. Front Plant Sci. 6. https://doi.org/10.3389/fpls.2015.00078.

Kumar, S., Singh, P., Khar, S., Sharma, M., 2016. Variability and association studies and screening of genotypes against pea seed borne mosaic virus (psmv) in lentil (*Lens culinaris* Medik.) under NWHimalayas of Jammu and Kashmir. Legum. Res. 39, 26–30.

Kwak, M., Velasco, D., Gepts, P., 2008. Mapping homologous sequences for determinacy and photoperiod sensitivity in common bean (*Phaseolus vulgaris*). J. Hered. 99, 283–291.

Lichtenzveig, J., David, J., Bon, W., Zhang, H.-B., Shtienberg, D., Abbo, S., 2006. Mapping quantitative trait loci in chickpea associated with time to flowering and resistance to *Didymella rabiei* the causal agent of Ascochyta blight. Theor. Appl. Genet. 113, 1357–1369.

Liu, Y., Gai, J.Y., Lü, H.N., Wang, Y.J., Chen, S.Y., 2005. Identification of drought tolerant germplasm and inheritance and QTL mapping of related root traits in soybean [(*Glycine max* (L.) Merr.)]. Yi Chuan Xue Bao 32, 855–863.

Liu, R., Jia, H., Cao, X., Huang, J., Li, F., Tao, Y., Qui, F., Zheng, Y., Zhang, Z., 2012. Fine mapping and candidate gene prediction of a pleiotropic quantitative trait locus for yield-related trait in *Zea mays*. PLoS ONE 7, e49836.

Lorenz, A.J., et al., 2011. Genomic selection in plant breeding. Adv. Agron. 110, 77–123.

Luby, J.J., Shaw, D.V., 2009. Plant breeders' perspectives on improving yield and quality traits in horticultural food crops. Horticult. Sci. 44 (1).

Luo, Y., Ma, T., Zhang, A., Ong, K.H., Li, Z., Yang, J., Yin, Z., 2016. Marker-assisted breeding of the rice restorer line Wanhui 6725 for disease resistance, submergence tolerance and aromatic fragrance. Rice 9. https://doi.org/10.1186/s12284-016-0139-9.

Michel, S., Ametz, C., Gungor, H., et al., 2016. Genomic selection across multiple breeding cycles in applied bread wheat breeding. Theor. Appl. Genet. 129, 1179–1189.

Muehlbauer, F.J., Weeden, N.F., Hoffman, D.L., 1989. Inheritance and linkage relationships of morphological and isozyme loci in lentil (*Lens* Miller). J. Hered. 80, 298–303.

Muehlbauer, F.J., Kasier, W.J., Clement, S.L., Summerfield, R.J., 1995. Production and breeding of lentil. Adv. Agron. 54, 283–332.

Nandan, R., Pandya, B.P., 1980. Correlation, path coefficient and selection indices in lentil. Indian J. Genet. Plant Breed. 40, 399–404.

Nelson, R.R., 1978. Genetics of horizontal resistance to plant diseases. Annu. Rev. Phytopathol. 16, 359–378.

Openshaw, S.J., Hadley, H.H., 1984. Selection indexes to modify protein concentration of soybean seeds. Crop Sci. 24, 1–4.

Pandey, S., Bhatore, A., Babbar, A., 2015. Studies on genetic variability, interrelationships association and path analysis in indigenous germplasm of Lentil in Madhya Pradesh, India. Electr. J. Plant Breed. 6, 592–599.

Paterson, A.H., Lander, E.S., Hewitt, J.D., Peterson, S., Lincoln, S.E., Tanksley, S.D., 1988. Resolution of quantitative traits into Mendelian factors by using a complete linkage map of restriction fragment length polymorphism. Nature 335, 721–726.

Paterson, A.H., Tanksley, S.D., Sorrells, M.E., 1991. DNA markers in plant improvement. Adv. Agron. 46, 39–90.

Pérez-Vega, E., Pañeda, A., Rodríguez-Suárez, C., Campa, A., Giraldez, R., Ferreira, J.J., 2010. Mapping of QTLs for morpho-agronomic and seed quality traits in a RIL population of common bean (*Phaseolus vulgaris* L.). Theor. Appl. Genet. 120, 1367–1380.

Phan, H.T.T., Ellwood, S.R., Hane, J.K., Ford, R., Materne, M., Oliver, R.P., 2007. Extensive macrosynteny between *Medicago truncatula* and *Lens culinaris* ssp. *culinaris*. Theor. Appl. Genet. 114, 549–558.

Rebetzke, G.J., van Herwaarden, A.F., Jenkins, C., Weiss, M., Lewis, D., Ruuska, S., Tabe, L., Fettell, N.A., Richards, R.A., 2008. Quantitative trait loci for water-soluble carbohydrates and associations with agronomic traits in wheat. Aust. J. Agric. Res. 59, 891–905.

Ribaut, J.M., de Vicente, M.C., Delannay, X., 2010. Molecular breeding in developing countries: challenges and perspectives. Curr. Opin. Plant Biol. 13, 213–218.

Richardson, K.L., Vales, M.I., Kling, J.G., Mundt, C.C., Hayes, P.M., 2006. Pyramiding and dissecting disease resistance QTL to barley stripe rust. Theor. Appl. Genet. 113, 485–495.

Rubeena, Ford, R., Taylor, P.W.J., 2003. Construction of an intraspecific linkage map of lentil (*Lens culinaris ssp. culinaris*). Theor. Appl. Genet. 107, 910–916.

Rubeena, Taylor, P.W.J., Ades, P.K., Ford, R., 2006. QTL mapping of resistance in lentil (*Lens culinaris ssp culinaris*) to ascochyta blight (*Ascochyta lentis*). Plant Breed. 125, 506–512.

Saha, G.C., Sarker, A., Chen, W., Vandemark, G.J., Muehlbauer, F.J., 2010a. Identification of markers associated with genes for rust resistance in *Lens culinaris* Medik. Euphytica 175, 261–265.

Saha, G.C., Sarker, A., Chen, W., Vandemark, G.J., Muehlbauer, F.J., 2010b. Inheritance and linkage map positions of genes conferring resistance to stemphylium blight in lentil. Crop Sci. 50, 1831–1839.

Saha, G.C., Sarker, A., Chen, W., Vandemark, G.J., Muehlbauer, F.J., 2013. Inheritance and linkage map positions of genes conferring agromorphological traits in *Lens culinaris* Medik. Int. J. Agron. 9. https://doi.org/10.1155/2013/618926.

Salehi, M., Haghnazari, A., Shakari, F., Faramarzi, F., 2008. The study of seed yield and seed yield components of lentil under normal and drought stress conditions. Pak. J. Biol. Sci. 11, 758–762.

Sandhu, J.S., Singh, S., 2007. History and origin. In: Yadav, S.S., McNeil, D.L., Stevenson, P.C. (Eds.), Lentil: An Ancient Crop for Modern Times. Springer, Dordecht, The Netherlands, pp. 1–9.

Sarker, A., Erskine, W., Sharma, B., Tyagi, M.C., 1999. Inheritance and linkage relationships of days to flower and morphological loci in lentil (*Lens culinaris* Medikus subsp. *culinaris*). J. Hered. 90, 270–275.

Schulthess, A.W., Wang, Y., Miedaner, T., Wilde, P., Reif, J.C., Zhao, Y., 2016. Multiple trait and selection indices genomic predictions for grain yield and protein content in rye for feeding purposes. Theor. Appl. Genet. 129, 273–287.

Servin, B., Martin, O.C., Mezard, M., Hospital, F., 2004. Toward a theory of marker-assisted gene pyramiding. Genetics 168, 513–523.

Sharma, S.K., Knox, M.R., Ellis, T.H.N., 1996. AFLP analysis of the diversity and phylogeny of *Lens* and its comparison with RAPD analysis. Theor. Appl. Genet. 93, 751–758.

Sharpe, A., Ramsay, L., Sanderson, L.A., Fedoruk, M., Clarke, W., Li, R., Kagale, S., Vijayan, P., Vandenberg, A., Bett, K., 2013. Ancient crop joins modern era: gene-based SNP discovery and mapping in lentil. BMC Genom. 14, 192–197.

Singh, S., Singh, I., Gill, R.K., Kumar, S., Sarker, A., 2009. Genetic studies for yield and component characters in large seeded exotic lines of Lentil. J. Food Legum. 22, 229–232.

Singh, D., Singh, C.K., Taunk, J., Tomar, R.S.S., 2016. Genetic analysis and molecular mapping of seedling survival drought tolerance gene in lentil (*Lens culinaris* Medikus). Mol. Breed. 36 (58). https://doi.org/10.1007/s11032-016-0474-y.

Smith, J.S.C., Smith, O.S., Wright, S., Wall, S.J., Walton, M., 1992. Diversity of U.S. hybrid maize germplasm as revealed by restriction fragment length polymorphisms. Crop Sci. 32, 598–604.

Steele, K.A., Price, A.H., Witcombe, J.R., Shrestha, R., Singh, B.N., Gibbons, J.M., Virk, D.S., 2012. QTLs associated with root traits increase yield in upland rice when transferred through marker-assisted selection. Theor. Appl. Genet. 126, 101–108.

Stuber, C.W., Moll, R.H., 1972. Frequency changes of isozyme alleles in a selection experiment for grain yield in maize (*Zea mays* L.). Crop Sci. 12, 337–340.

Stuber, C.W., Goodman, M.M., Moll, R.H., 1982. Improvement of yield and ear number resulting from selection at allozyme in a maize population. Crop Sci. 22, 737–740.

Sudheesh, S., Rodda, M.S., Davidson, J., Javid, M., Stephens, A., Slater, A.T., Noel, O.I., Cogan, N.O.I., Forster, J.W., Kaur, S., 2016. SNP-based linkage mapping for validation of QTLs for resistance to ascochyta blight in lentil. Front. Plant Sci. 7, 1604. http://doi.org/10.3389/fpls.2016.01604.

Suwantaradon, K., Eberhart, S.A., Mock, J.J., Owens, J.C., Guthrie, W.D., 1975. Index selection for several agronomic traits in the BSSS2 maize population. Crop Sci. 15, 827–833.

Tadmor, Y., Zamir, D., Ladizinsky, G., 1987. Genetic mapping of an ancient translocation in the genus *Lens*. Theor. Appl. Genet. 73, 883–892.

Tahir, M., Muehlbauer, F.J., 1994. Gene mapping in lentil with recombinant inbred lines. J. Hered. 85, 306–310.

Tahir, M., Simon, C.J., Muehlbauer, F.J., 1993. Gene map of lentil: a review. LENS Newsl. 20, 3–10.

Tahir, M., Lindeboom, N., Baga, M., Vandenberg, A., Chibbar, R.N., 2011. Composition and correlation between major seed constituents in selected lentil (*Lens culinaris* Medik.) genotypes. Can. J. Plant Sci. 91, 825–835.

Taiery, A., Mirshekari, B., 2014. Determination of some attributes of lentil (*Lens culinaris*) with higher effect on yield via step by step regression analyses. Int. J. Biosci. 4, 170–174.

Tanksley, S.D., 1993. Mapping polygenes. Ann. Rev. Genet. 27, 205–233.

Tanksley, S.D., Medina, H.F., Rick, C.M., 1982. Use of naturally occurring enzyme variation to detect and map genes controlling quantitative traits in an interspecific backcross of tomato. Heredity 49, 11–25.

Tanyolac, B., Ozatay, S., Kahraman, A., Muehlbauer, F.J., 2010. Linkage mapping of lentil (*Lens culinaris* L.) genome using recombinant inbred lines revealed by AFLP, ISSR, RAPD and some morphologic markers. J. Agric. Biotechnol. Sustain. Dev. 2, 001–006.

Tarán, B., Buchwaldt, L., Tullu, A., Banniza, S., Warkentin, T.D., Vandenberg, A., 2003. Using molecular markers to pyramid genes for resistance to ascochyta blight and anthracnose in lentil (*Lens culinaris* Medik.). Euphytica 134 (2), 223–230.

Tayeh, N., Klein, A., Le Paslier, M.C., et al., 2015. Genomic prediction in pea: effect of marker density and training population size and composition on prediction accuracy. Front Plant Sci. 6, 1–11.

Temel, H.Y., Gol, D., Kahriman, A., Tanyolac, M.B., 2014. Single nucleotide polymorphism discovery through Illumina-based transcriptome sequencing and mapping in lentil. Turk. J. Agric. Forestry 38, 1–19.

Thoday, J.M., 1961. Location of polygenes. Nature 191, 368–370.

Tuba, B.B., Sakar, D., 2008. Studies on variability of lentil genotypes in South-eastern anatolia of Turkey. Notulae Bot. Horti Agrobot. Çluj-Napoca 36, 20–24.

Tullu, A., Buchwaldt, L., Warkentin, T., Taran, B., Vandenberg, A., 2003. Genetics of resistance to anthracnose and identification of AFLP and RAPD markers linked to the resistance gene in PI 320937 germplasm of lentil (*Lens culinaris* Medikus). Theor. Appl. Genet. 106, 428–434.

Tullu, A., Tar'an, B., Breitkreutz, C., Banniza, S., Warkentin, T.D., Vandenberg, A., Buchwaldt, L., 2006. A quantitative-trait locus for resistance to ascochyta blight [*Ascochyta lentis*] maps close to a gene for resistance to anthracnose [*Colletotrichum truncatum*] in lentil. Can. J. Plant Pathol. 28, 588–595. http://doi.org/10.1080/07060660609507337.

Tullu, A., Tar'an, B., Warkentin, T., Vandenberg, A., 2008. Construction of an intraspecific linkage map and QTL analysis for earliness and plant height in lentil. Crop Sci. 48, 2254–2264.

Tyagi, S.D., Khan, M.H., 2010. Studies on genetic variability and interrelationship among the different traits in Microsperma lentil (*Lensculinaris* Medik.). J. Agric. Biotechnol. Sustain. Dev. 2, 015–020.

Vaillancourt, R.E., Slinkard, A.E., 1993. Linkage of morphological and isozyme loci in lentil, *Lens culinaris* L. Can. J. Plant Sci. 73, 917–926.

Varshney, R.K., Tuberosa, R., 2007. Genomics-assisted crop improvement: an overview. In: Varshney, R.K., Tuberosa, R. (Eds.), Genomics-Assisted Crop Improvement, vol 1: Genomics Approaches and Platforms. Springer, Dordrecht, The Netherlands, pp. 1–12.

Verma, P., Sharma, T.R., Srivastava, P.S., Abdin, M.Z., Bhatia, S., 2014. Exploring genetic variability within lentil (*Lens culinaris* Medik.) and across related legumes using a newly developed set of microsatellite markers. Mol. Biol. Rep. 41, 5607–5625.

Verma, P., Goyal, R., Chahota, R.K., Sharma, T.R., Abdin, M.Z., Bhatia, S., 2015. Construction of a genetic linkage map and identification of QTLs for seed weight and seed size traits in lentil (*Lens culinaris* Medik.). PLoS ONE 10 (10), e0139666.

Vikram, P., Swamy, B.P.M., Dixit, S., Ahmed, H.U., Cruz, M.T.S., Singh, A.K., Kumar, A., 2011. $qDTY_{1.1}$, a major QTL for rice grain yield under reproductive-stage drought stress with a consistent effect in multiple elite genetic backgrounds. BMC Genet. 12, 89–95.

Wang, X., Jiang, G.L., Green, M., Scott, R.A., Hyten, D.L., Cregan, P.B., 2012. Quantitative trait locus analysis of saturated fatty acids in a population of recombinant inbred lines of soybean. Mol. Breed. https://doi.org/10.1007/s11032-012-9704-0.

Weeden, N.F., Muehlbauer, F.J., Ladizinsky, G., 1992. Extensive conservation of linkage relationships between pea and lentil genetic maps. J. Hered. 83, 123–129.

Wei, D., Cui, K., Ye, G., Pan, J., Xiang, J., Huang, J., Nie, L., 2012. QTL mapping for nitrogen-use efficiency and nitrogen-deficiency tolerance traits in rice. Plant Soil 359, 281–295.

Wong, M.M.L., Verma, N.G., Ramsay, L., Yuan, H.Y., Caron, C., Diapari, M., Vandenberg, A., Bett, K.E., 2015. Classification and characterization of species within the genus *Lens* using genotyping-by-sequencing (GBS). PLoS ONE 10 (3), e0122025. https://doi.org/10.1371/journal.pone.0122025.

Worland, A.J., 1996. The influence of flowering time genes on environmental adaptability in European wheats. Euphytica 89, 49–57.

Xu, Y., Crouch, J.H., 2008. Marker-assisted selection in plant breeding: from publications to practice. Crop Sci. 48, 391–407.

Xu, Y., et al., 2012. Whole-genome strategies for marker-assisted plant breeding. Mol. Breed. 29, 833–854.

Ye, G., Smith, K.F., 2008. Marker-assisted gene pyramiding for inbred line development: practical applications. Int. J. Plant Breed. 1 (2), 11–22.

Zamir, D., Ladizinsky, G., 1984. Genetics of allozyme variants and linkage groups in lentil. Euphytica 33, 329–336.

Závodná, M., Kraic, J., Paglia, G., Gregova, E., Morgante, M., 2000. Differentiation between closely related lentil (*Lens culinaris* Medik.) cultivars using DNA markers. Seed Sci. Technol. 28, 217–219.

Zhang, W.B., Qiu, P.C., Jiang, H.W., Liu, C.Y., Xin da, W., Li, C.D., Hu, G.H., Chen, Q.S., 2012. Dissection of genetic overlap of drought and low-temperature tolerance QTLs at the germination stage using backcross introgression lines in soybean. Mol. Biol. Rep. 39, 6087–6094.

FURTHER READING

Chen, S.Y., 2005. Identification of drought tolerant germplasm and inheritance and QTL mapping of related root traits in soybean [(*Glycine max* (L.) Merr.)]. J. Genet. Genom. 32, 855–863.

Ford, M.A., Austin, R.B., Angus, W.J., Sage, G.C.M., 1981. Relationships between responses of spring wheat genotypes to temperatures photoperiodic treatments and their performance in the field. J. Agric. Sci. 96, 623–634.

Chapter 7

Molecular Marker Assisted Gene Pyramiding

Jitendra Kumar*, Shiv Kumar†, Debjyoti Sen Gupta*, Sonali Dubey*, Sunanda Gupta*, Priyanka Gupta*

**Division of Crop Improvement, Indian Institute of Pulses Research, Kanpur, India, †Biodiversity and Integrated Gene Management Program, International Center for Agricultural Research in the Dry Areas (ICARDA), Rabat, Morocco*

7.1. INTRODUCTION

The lentil (*Lens culinaris* ssp. *culinaris* Medik.) is an important cool-season crop. It is a diploid ($2n = 2\times = 14$), self-pollinated crop and has a genome size of 4 Gbp (Arumuganathan and Earle, 1991). It is a rich source of dietary proteins (22%–35%), minerals, fiber, and carbohydrates. Consumption of 50 g of lentil per day helps to alleviate malnutrition, especially in developing countries. Lentil carbohydrates have a low glycemic index, so they are highly recommended by physicians for consumption by people suffering from diabetes, obesity, and cardiovascular diseases.

The production and productivity of this crop are still low due to the lentil's poor ability to compete with weeds, its higher flower-drop rate, tendency toward pod shedding, and several other biotic and abiotic factors (Sharpe et al., 2013). Further, in recent years, sporadic changes in climatic conditions have begun to affect production and productivity in lentil cultivation.

During the past years, systematic and continuous breeding efforts have been made through conventional methods for improving productivity. However, we could not achieve substantial genetic gain, and this is reflected in the progress made in lentil productivity in India, which increased only marginally from 539 kg/ha in 1960–61 to 759 kg/ha in 2013–14 (AICRP on MULLaRP, 2014–15). One of the major reasons for this slow gain is high genotype × environment (G × E) interactions on the expression of important quantitative traits (Kumar and Ali, 2006). Therefore, recently, attention has been focused on strengthening conventional breeding programs by using new breeding tools. Now, deployment of molecular markers is routinely used world-wide in all major cereal crops and in major pulses like the common bean, chickpeas, and pigeon peas (Gupta et al., 2010; Kumar et al., 2017). In the past few years, major emphasis has also

Lentils. https://doi.org/10.1016/B978-0-12-813522-8.00007-8

been placed on developing similar kinds of genomic resources for improving the productivity of lentil crops (Varshney et al., 2009, 2010; Sato et al., 2010). In lentil marker-trait association, analysis has been established for several agronomically important traits. Use of linked molecular markers can help to provide high-yielding genotypes by improving those traits, which are otherwise difficult to improve through conventional breeding methods. This chapter discusses the present use of marker-assisted selection in lentils.

7.2. MOLECULAR MARKERS

For marker-assisted selection, the variability of molecular markers is an essential requirement for developing linkage maps and identifying the markers' close association with trait of interest. In lentils, various molecular markers such as RAPD, SSR, AFLP, and RFLP markers were developed and used in analysis of genetic diversity and construction of linkage maps (Havey and Muehlbauer, 1989, Weeden et al., 1992; Eujayl et al., 1998a). Among the various PCR-based markers, SSR markers have made significant contribution to the recent development of lentil genome maps. These markers have been developed and used in lentils (Hamwieh et al., 2009). The express-sequenced tag provided an opportunity for developing SSR and SNP markers. Therefore, Expressed Sequenced Tags (EST) have been generated through cDNA libraries in lentils (Kumar et al., 2014) and are now available in the public domain. These ESTs have led to the easy, cost-effective development of SSR and SNP markers using some special bioinformatic programs, for example, MISA for SSR detection (Thiel et al., 2003) and SNipPer for SNP discovery (Kota et al., 2003).

In recent years, second-generation technology has begun carrying out transcriptome sequencing of lentils, which permitted large-scale unigene assembly and SSR marker discovery (Kaur et al., 2011). Researchers developed a set of 2393 EST-SSR markers developed in lentils, and a subset of 192 EST-SSR markers have been validated across a panel of 12 cultivars with 47.5% polymorphism (Kaur et al., 2011). Recently, transcriptome cDNA library sequencing using the Illumina GA/GAIIx system has provided a potential alternative. As a result of transcriptome sequencing, massive data was obtained in the form of about 847,824 high-quality sequence reads and transcriptome assemblies with 84,074 unigenes.

In recent years, next-generation sequencing approaches have quickly replaced SSR markers with DNA chip-based SNP markers. About 44,879 SNP markers have been identified in lentils using the Illumina Genome Analyzer (Sharpe et al., 2013). The recent discovery of high-density SNP markers has facilitated the establishment of ultra HTP genotyping technologies such as Illumina GoldenGate (GG), which can accommodate >1000 SNPs in GG platforms (Kaur et al., 2014; Sharpe et al., 2013). Because SNP discovery and genotyping require expensive and sophisticated platforms, the development and exploitation of SNP markers is still limited in lentils. There are techniques available to detect SNPs such as allele-specific PCR, single-base extension, and array hybridization methods. They are cost-effective methods, and through the use of allele-specific PCR (KASPar) markers, we can include a small-to-moderate number of SNPs for any specific application (Fedoruk et al., 2013; Sharpe et al., 2013).

7.3. TRAIT-SPECIFIC MAPPING POPULATIONS

For establishing the association of marker with a trait of agronomic importance, we need bi-parental or multiparental mapping populations or a diverse panel of genotypes for association mapping. Biparental mapping populations include F_2, backcross, double haploid (DH), and recombinant inbred line (RIL) populations. In the lentil, different institutions tried to develop biparental mapping populations for key traits and used them in marker trait association analysis (Table 7.1). RIL populations were developed from crosses made between contrasting parents for the traits of interest through the single-seed descent method. Indian Institute of Pulses Research (IIPR) has recently developed RIL population from a cross between ILL6002 and ILL7663 in order to identify and map early growth vigor genes in lentils. Identification of markers linked to the genes/QTLs governing these traits will help develop a genotype having high biomass at an early stage. Similarly, for tagging and mapping of genes of earliness,

TABLE 7.1 RIL Mapping Populations Developed for Various Traits in Lentils

Trait	Cross	Population Size	Institute
Drought	ILL7946 × ILL7979	174	ICARDA
Cold	ILL4605 × ILL10657	153	ICARDA
Earliness	ILL7115 × ILL8009	150	ICARDA
Rust	ILL5888 × ILL6002	152	ICARDA
Fusarium wilt	ILL213 × ILL5883, Precoz × Idleb 2	150	ICARDA
Zinc content	ILL5722 × ILL9888	177	ICARDA
	ILL9888 × ILL5480	149	ICARDA
Iron content	ILL9932 × ILL9951	193	ICARDA
Early growth vigor	DPL 15 × ILL7663	160	IIPR, Kanpur
Root traits	IPL 98/193 × EC208362	160	IIPR, Kanpur
Earliness	Precoz × L 4603	160	IIPR, Kanpur

another mapping population was developed from a cross between Precoz (medium early duration) and L4603 (early duration). CSK Himachal Pradesh Agricultural University in Palampur, India, has developed recombinant inbred populations involving both intra- and intersubspecific crosses where parents differ for rust reaction, drought tolerance, flowering time, plant vigor, shattering tolerance, seed size, and seed weight. With rapid generation advancement technology (Mobini et al., 2014), which allows 4–5 generations per year in lentils, researchers will boost the development of much-needed genetic resources for genomics-enabled improvement.

7.4. DEVELOPMENT OF MOLECULAR LINKAGE MAPS

The first genetic mapping (linkage analysis) began in lentils in 1984 (Ladizinsky, 1984), and the first map comprising DNA-based markers was produced by Havey and Muehlbauer (1989). Subsequent maps were published by several researchers (Table 7.2). With the development of PCR-based markers, the number of mapped markers across the *Lens* genome increased dramatically (Kumar et al., 2011, 2014). The first extensive map comprised of RAPD, AFLP, RFLP, and morphological markers was constructed using a RIL population from a cross between a cultivated *L. culinaris* ssp. *culinaris* cultivar and a *L. culinaris* ssp. *orientalis* accession (Eujayl et al., 1998a). Rubeena et al. (2003) published the first intraspecific lentil map comprising 114 RAPD, intersimple sequence repeat (ISSR), and resistance gene analogue (RGA) markers. Rubeena et al. (2006) reported an F_2 map comprising 72 markers (38 RAPD, 30 AFLP, 3 ISSR, and one morphological marker) spanning 412.5 cM. The first *Lens* map to include SSR markers was that of Duran et al. (2004). Hamwieh et al. (2005) added 39 SSR and 50 AFLP markers to the map constructed by Eujayl et al. (1998a) to produce a comprehensive *Lens* map comprising 283 genetic markers covering 715 cM. Subsequently, the first lentil map that contained 18 SSR and 79 cross genera ITAP gene-based markers was constructed using a F_5 RIL population developed from a cross between ILL5722 and ILL5588 (Phan et al., 2007). The map comprised seven linkage groups that varied from 80.2 to 274.6 cM in length and spanned a total of 928.4 cM (Gupta et al., 2012a) used 196 markers including new 15 *Medicago truncatula* EST-SSR/SSR using a population of 94 RIL produced from a cross between ILL5588 and ILL5722 and clustered into 11 linkage groups (LG) covering 1156.4 cM. An inter-subspecific F_2 *Lens* linkage map consisting of 199 PCR-based markers (28 SSRs, 9 ISSRs and 162 RAPDs) mapped onto 11 linkage groups covering a distance of 3847 cM has been constructed (Gupta et al., 2012b). The population-specific linkage maps have been developed by Andeden et al. (2013), Ford et al. (2007), and Perez de la Vega et al. (2011). More recently, a map was developed from an interspecific population based on 377 SNP markers. These SNP markers have been developed using a GoldenGate genotyping array and single SNP marker assays (Gujaria-Verma et al., 2014). More recently, an intraspecific linkage map of

TABLE 7.2 List of Molecular Linkage Maps Developed in Lentils

Crossing Species	Type and Size (in Parenthesis) of Population	Type of Marker Mapped	Mapped Length (cM) and No. of Loci (in Parenthesis)	References
Lens culinaris ssp. *culinaris* × *L. c.* ssp. *orientalis* crosses	RIL (14–80)	Isozyme and four morphological markers	– (20)	Tahir and Muehlbauer (1994)
Lens culinaris ssp. *culinaris* × *L. c.* ssp. *orientalis* crosses	F_2	Isozymes	– (10)	Zamir and Ladizinsky (1984)
Lens culinaris × *L. ervoides* and *Lens culinaris* × *L. ervoides* crosses	F_3 (107) F_3 (22–56)	Isozymes	258 (18)	Tadmor et al. (1987)
Lens culinaris ssp. *orientalis* × *Lens culinaris* ssp. *Culinaris*	RIL (86)	RAPD, AFLP, RFLP, and morphological markers	1073 (177)	Eujayl et al. (1998a)
Lens culinaris ssp. *Culinaris* × *L. c.* ssp. *Orientalis*	F_2 (113)	RAPD, ISSR, AFLP, SSR, CAPS, SRAPS, and morphological markers	2234 (200)	Duran et al. (2004), Fratini et al. (2004), de la Puente et al. (2013)
ILL5588 × L692–16-1 (s)	RIL (86)	SSR, AFLP	751 (283)	Hamwieh et al. (2005)
ILL5588 × ILL7537	F_2 (150)	RAPD, ISSR and RGA	784 (114)	Rubeena et al. (2003)
Eston × PI 320937	RIL (94)	AFLP, RAPD, and SSR	1868 (207)	Tullu et al. (2006, 2008)
Precoz × WA 8649041	RIL (94)	AFLP, ISSR, RAPD, and morphological markers	1396 (166)	Tanyolac et al. (2010)
ILL 6002 × ILL 5888	RIL (206)	SSR, RAPD, SRAP, and morphological markers	1565 (139)	Saha et al. (2010a, 2013)

Continued

TABLE 7.2 List of Molecular Linkage Maps Developed in Lentils—cont'd

Crossing Species	Type and Size (in Parenthesis) of Population	Type of Marker Mapped	Mapped Length (cM) and No. of Loci (in Parenthesis)	References
WA8649090 × Precoz	RIl (106)	RAPD, ISSR, and AFLP	1192 (130)	Kahraman et al. (2004, 2010)
ILL5722 × ILL5588	RIL (94)	RAPD, ISSR, ITAP, and SSR	1392 (211)	Gupta et al. (2012a)
L830 × ILWL77	F_2 (114)	SSR, ISSR, and RAPD	3843 (199)	Gupta et al. (2012b)
CDC Robin × 964a-46	RIL (139)	SNP, SSR, and seed color genes	697 (561)	Sharpe et al. (2013) and Fedoruk et al. (2013)
Cassab × ILL 2024	RIL (126)	SSR and SNP	1178 (318)	Kaur et al. (2014)
PI 320937 × Eston	RIL (96)	AFLP, SSR, and SNP	840 (194)	Sever et al. (2014)
Precoz × WA 8649041	RIL (101)	SNP	540 (519)	Temel et al. (2014)
ILL 8006 × CDC Milestone	–	AFLP, SSR, and SNP	497 (149)	Aldemir et al. (2014)
Precoz × L830	RIL (126)	SSR	1184 (219)	Verma et al. (2015)
Indianhead × Northfield; Indianhead × Digger; Northfield × Digger	RILs (117, 112, 114)	SNP	2429.6 (689)	Sudheesh et al. (2016)
L01-827A (*L. ervoides*) × IG 72815 (*L. ervoides*)	RIL (94)	SNP	740.9 (543)	Bhadauria et al. (2017)
ILL 8006 × "CDC Milestone"	RIL (118)	SNP	497.1 (4177)	Aldemir et al. (2017)

Source: This table is updated from Kumar, S., Rajendran, K., Kumar, J., Hamwieh, A., Baum, M., 2015. Current knowledge in lentil genomics and its application for crop improvement. Front. Plant Sci. 6, 78.

216 loci was constructed using three mapping populations. This map spanned a 1183.7 cM distance of the lentil genome with an average marker density of 5.48 cM (Verma et al., 2015). The list of comprehensive maps in lentils are provided with details in Table 7.2.

7.5. MARKER-TRAIT LINKAGE ANALYSIS (MAPPING OF GENES/QTL) USING BIPARENTAL POPULATIONS

In lentils, molecular markers linked to desirable genes/QTL have been identified during the past several years which have been listed in Table 7.3. However, these markers were not used in marker-assisted selection in lentils for several reasons, including poor linkage between markers and traits and low phenotypic variation explained. For example, the association between DNA markers and the *Fusarium* wilt-resistance (Fw) gene was confirmed by several researchers (Eujayl et al., 1998b; Hamwieh et al., 2005), but could not be used in marker-assisted selection due to the fact that the SSR59-2B marker was loosely linked with Fw at a distance of 19.7 cM (Hamwieh et al., 2005). Similarly, for several other traits, flanking markers were identified as being located at more than a 10 cM distance. Although these markers have explained high phenotypic variance, these markers cannot be used in a marker-assisted breeding program. For these QTLs, fine mapping is still required to make them useful for crop improvement in lentils. However, Verma et al. (2015) identified QTLs for the seed weight (*qSW*) and seed size (*qSS*) traits that were co-localized on LG4 and explained 48.4% and 27.5% of phenotypic variance, respectively.

7.6. MARKER-TRAIT ASSOCIATION ANALYSIS USING NATURAL DIVERSE POPULATION

Though, biparental mapping has resulted in the identification of a number of markers associated with genes/QTL governing traits of interest, their use in marker-assisted breeding programs is still limited because parents used for making genetic improvement might not be different for identified, targeted QTLs (Bernardo, 2008). Moreover, identification of QTLs for traits based on a biparental mapping population has also been discouraged due to poor maker-trait association and the inefficient, costly process of fine mapping (Parisseaux and Bernardo, 2004). Alternatively, association mapping (AM), which uses a set of diverse genotypes derived from known/unknown ancestry for genes/QTLs mapping, helped to overcome the disadvantage of bi-parental mapping. In association mapping, an association panel comprised of natural germplasm lines, breeding lines developed in breeding programs, individuals of multiple biparental populations, and/or populations developed by crossing multiple parents are used (Steinhoff et al., 2011; Wurschum, 2012; Gupta et al., 2014). This is cost-effective approach for QTL because phenotypic data recorded routinely on breeding lines can be used for QTL mapping. Use of diverse breeding lines

TABLE 7.3 Molecular Markers Linked to Genes/QTLs Controlling Important Genetic Distance of Flanking Markers and Phenotypic Variance Explained (PVE) by Linked QTLs in Lentils

Traits	Type of Linked Markers	Distance of Flanking Marker From Locus (cM)	PVE (%)	References
Ascophyta blight resistance	RAPD	6–14	90	Ford et al. (1999)
	RAPD, AFLP, ISSR	–	~50	Rubeena et al. (2006)
	AFLP, RAPD	7	41	Tullu et al. (2006)
	ITAP, SSR, ISSR	–	~61	Gupta et al. (2012a)
	SNP and SSR	3.5	52–69	Sudheesh et al. (2016)
Anthracnose	SNP	–	8.9–24.8	Bhadauria et al. (2017)
Earliness	RAPD, AFLP, and SSR	–	37–46	Tullu et al. (2008)
Plant height	RAPD, AFLP, and SSR		31–40	Tullu et al. (2008)
	RAPD, AFLP, and SSR and morphological markers	12–36	38.2	Fratini et al. (2007)
Branches at the first node	RAPD, ISSR, AFLP and SSR and morphological markers	12–14	91.7	Fratini et al. (2007)
Total no. of branches	RAPD, ISSR, AFLP, and SSR and morphological markers	46	54	Fratini et al. (2007)
Height at the first node	RAPD, ISSR, AFLP, and SSR and morphological markers	26–28	33.3	Fratini et al. (2007)
Flowering time	RAPD, ISSR, AFLP, and SSR and morphological markers	6–26	90.4	Fratini et al. (2007)

Pod dehiscence	RAPD, ISSR, AFLP, and SSR and morphological markers	14–48	81.3	Fratini et al. (2007)
Seed weight	morphological markers	10–12	18.2	Fratini et al. (2007)
Seed diameter	RAPD, ISSR, AFLP, and SSR and morphological markers	28–33	37	Fratini et al. (2007)
Winter hardiness	RAPD, ISSR, and AFLP	1.2	20.45	Kahraman et al. (2010)
Cotyledon color class (Yc)	SNP, SSR, and seed color loci	23	23	Fedoruk et al. (2013)
Seed thickness	SNP, SSR, and seed color loci	3.7–71.4	84	Fedoruk et al. (2013)
Seed diameter	SNP, SSR, and seed color loci	7.5–59.7	~60	Fedoruk et al. (2013)
Seed plumpness	SNP, SSR, and seed color loci	7.5–59.7	~50	Fedoruk et al. (2013)
Days to 50% flowering	SNP, SSR, and seed color loci	5.4–76.9	~34	Fedoruk et al. (2013)
	SSR, SRAP, RAPD	1.4–7.2	24.2	Saha et al. (2013)
Hundred-seed weight	SSR, SRAP, RAPD	0.4–7	17.5	Saha et al. (2013)
Plant height	SSR, SRAP, RAPD	4	15.3	Saha et al. (2013)
Seed diameter	SSR, SRAP, RAPD	0.4–2.6	32.6	Saha et al. (2013)
Stemphylium blight resistance	SSR, SRAP, RAPD	–	46	Saha et al. (2010a)
	SNP	–	9.9–18.3	Bhadauria et al. (2017)
Boron tolerance	EST-SNP	–	71	Kaur et al. (2014)
Iron concentration	SNP	0.12	5.9%–14.0%	Aldemir et al. (2017)

Source: This table is updated from Kumar, S., Rajendran, K., Kumar, J., Hamwieh, A., Baum, M., 2015. Current knowledge in lentil genomics and its application for crop improvement. Front. Plant Sci. 6, 78.

in association mapping is more useful compared to other kind of diverse panels (Gupta et al., 2014). Because breeding populations are generally narrow in their genetic base, QTLs identified in the background of such populations can be used directly in marker-assisted selection, while those QTLs identified from crosses made between cultivated and wild species are useful only for the introgression of traits from exotic material (Kumar et al., 2017). However, limited efforts were used to study the marker-trait association analysis in lentils. Fedoruk et al. (2013), first used AM in lentils to identify QTLs for seed size and seed shape. As the properly designed AM panels have a greater frequency of alleles, encompassing the genetic variation of a crop, it could save both time and cost when compared to MAS in lentils. More recently, the functional markers associated with flowering time in lentils were identified through association mapping (Kumar et al., 2018).

7.7. PRESENT STATUS AND FUTURE PROSPECTS OF MARKER-ASSISTED SELECTION

During the past several years, significant progress has been made towards the development of molecular markers in lentils. These markers can be used for evaluation of breeding materials and introgression of QTLs/genes for economically important traits through marker-assisted backcrossing. In the following text, the present status and future prospects of using molecular markers in lentil breeding programs have been discussed.

7.7.1 Marker-Assisted Evaluation of Breeding Materials

Knowledge of the genetic diversity available in elite gene pools is required for selecting diverse parental lines for hybridization. This helps to broaden the genetic base of core breeding materials for selecting useful recombinants. In the lentil, during the past several years, the genetic diversity of the lentil germplasm has been assessed preferably using molecular markers (Udupa et al., 1999; Abe et al., 2003; Hamwieh et al., 2009; Reddy et al., 2009; Dikshit et al., 2015; Idrissi et al., 2015; Khazaei et al., 2016; Kumar et al., 2016). However, these diversity analysis results could not be practically utilized for selecting parental lines in breeding programs. Use of molecular markers has helped to identify true F_1 plants. In a study, use of molecular markers in lentils detected only 21% of plants were true hybrids, which enhanced plant breeders' efficiency in selecting recombinant plants and saved the time and money required to grow a population from selfed or admixed plants (Solanki et al., 2010).

7.7.2 Marker-Assisted Backcrossing Program

Molecular markers linked to genes/QTL controlling a trait can be used to introgress that QTL in the background of improved lines through marker-assisted breeding.

In lentils, although molecular markers linked to desirable genes/QTL have been reported, only those with tight association (<1.0cM) and high phenotypic effect can be used in marker-assisted selection (Table 7.3). For example, mapping of *Ascochyta* blight resistance using an F_2 population derived from ILL7537×ILL6002 identified three QTLs accounting for 47% (*QTL*-1 and *QTL*-2) and 10% (*QTL*-3) of disease variation. Therefore, QTLs that explained >47% of total phenotypic variation can be used in marker-assisted breeding. Recently, QTLs conferring resistance to *Stemphylium* blight and rust diseases using RIL populations were identified in lentils (Saha et al., 2010a; Saha et al., 2010b). Though F_2 lentil populations have been used widely in identification of QTLs, their use in marker-trait analysis has led to identification of only major QTLs. Thus, several minor QTLs were overlooked in such populations, and identification of environmental-responsive QTLs was difficult. Because quantitative traits are influenced by both genetic and environmental effects, RILs or near isogenic lines (NILs) are more suitable populations to accurately dissect their components. For *Ascochyta* blight, three QTLs each were detected for resistance at seedling and pod/maturity stages (Gupta et al., 2012a). Together, these accounted for 34% and 61% of the total estimated phenotypic variation and demonstrated that resistance at different growth stages is potentially conditioned by different genomic regions. The flanking markers identified may be useful for MAS and pyramiding of potentially different resistance genes into elite backgrounds that are resistant throughout the cropping season. One QTL each for the seed weight (*qSW*) and seed size (*qSS*) traits, explaining 48.4% and 27.5% of phenotypic variance, respectively, were identified in lentils. These QTLs were located on average 5.48cM from their marker, indicating close marker-trait association, so they can be useful in marker-assisted breeding for improving seed size and weight (Verma et al., 2015). Therefore, for successful use of molecular markers in lentil breeding programs, there is a need to identify tightly linked molecular markers with genes/QTLs controlling important traits.

REFERENCES

Abe, J., Xu, D.H., Suzuki, Y., Kanazawa, A., Shimanmoto, Y., 2003. Soybean germplasm pools in Asia revealed by nuclear SSRs. Theor. Appl. Genet. 106, 445–453.

Aldemir, S.B., Sever, T., Ates, D., Yagmur, B., Kaya, H.B., Temel, H.Y., et al., 2014. QTL mapping of genes controlling Fe uptake in lentil (*Lens culinaris* L.) seed using recombinant inbred lines. In: Plant and Animal Geonme Conference. vol. XXII. 3360 p. https://pag.confex.com/pag/xxii/webprogram/Paper9689.html.

Aldemir, S., Ateş, D., Temel, H.Y., Yağmur, B., Alsaleh, A., Kahriman, A., Özkan, H., Vandenberg, A., Tanyolaç, M.B., 2017. QTLs for iron concentration in seeds of the cultivated lentil (*Lens culinaris* Medic.) via genotyping by sequencing. Turk. J. Agric. For. 41, https://doi.org/10.3906/tar-1610-33.

Andeden, E.E., Derya, M., Baloch, F.S., Kilian, B., Ozkan, H., 2013. In: Development of SSR markers in lentil. Proceedings of Plant and Animal Genome Conference XXI. 351 p. https://pag.confex.com/pag/xxi/webprogram/Paper6565.html.

Arumuganathan, K., Earle, E.D., 1991. Nuclear DNA content of some important plant species. Plant Mol. Biol. Rep. 9, 208–218.

Bernardo, R., 2008. Molecular markers and selection for complex traits in plants: Learning from the last 20 years. Crop Sci. 48, 1649–1664.

Bhadauria, V., Ramsay, L., Bett, K.E., Banniza, S., 2017. QTL mapping reveals genetic determinants of fungal disease resistance in the wild lentil species Lens ervoides. Sci. Rep. 7. https://doi.org/10.1038/s41598-017-03463-9.

de la Puente, R., Garcia, P., Polanco, C., Perez de la Vega, M., 2013. An improved intersubspecific genetic map in *Lens* including functional markers. Spanish J. Agric. Res. 11, 132–136.

Dikshit, H.K., Singh, A., Singh, D., Aski, M.S., Prakash, P., Jain, N., Meena, S., Kumar, S., Sarker, A., 2015. Genetic diversity in *Lens* species revealed by EST and genomic simple sequence repeat analysis. PLoS ONE 10, e0138101.

Duran, Y., Fratini, R., Garcia, P., De la Vega, M.P., 2004. An intersubspecific genetic map of Lens. Theor. Appl. Genet. 108, 1265–1273.

Eujayl, I., Baum, M., Powell, W., Erskine, W., Pehu, E., 1998a. A genetic linkage map of lentil (*Lens* sp.) based on RAPD and AFLP markers using recombinant inbred lines. Theor. Appl. Genet. 97, 83–89.

Eujayl, I., Erskine, W., Bayaa, B., Baum, M., Pehu, E., 1998b. Fusarium vascular wilt in lentil: Inheritance and identification of DNA markers for resistance. Plant Breed. 117, 497–499.

Fedoruk, M.J., Vandenberg, A., Bett, K.E., 2013. Quantitative trait loci analysis of seed quality characteristics in lentil using single nucleotide polymorphism markers. Plant Genome 6. https://doi.org/10.3835/plantgenome2013.05.0012.

Ford, R., Pang, E.C.K., Taylor, P.W.J., 1999. Genetics of resistance to Ascochyta blight (*A. lentis*) of lentil and the identification of closely linked RAPD markers. Theor. Appl. Genet. 98, 93–98.

Ford, R., Rubeena Redden, R.J., Materne, M., Taylor, P.W.J., 2007. Lentil. In: Kole, C. (Ed.), Genome Mapping and Molecular Breeding in Plants: Pulses, Sugar and Tuber Crops. Springer, Berlin, pp. 91–104.

Fratini, R., Ruiz, M.L., de la Vega, M.P., 2004. Intra-specific and inter-sub-specific crossing in lentil (*Lens culinaris* Medik.). Can. J. Plant Sci. 84, 981–986.

Fratini, R., Duran, Y., Garcia, P., Pérez de la Vega, M., 2007. Identification of quantitative trait loci (QTL) for plant structure, growth habit and yield in lentil. Span. J. Agric. Res. 5, 348–356.

Gujaria-Verma, N., Vail, S.L., Carrasquilla-Garcia, N., Penmetsa, R., Cook, D.R., Farmer, A.D., Vandenberg, A., Bett, K.E., 2014. Genetic mapping of legume orthologs reveals high conservation of synteny between lentil species and the sequenced genomes of Medicago and chickpea. Front. Plant Sci. 5, 676.

Gupta, P.K., Kumar, J., Mir, R.R., Kumar, A., 2010. Marker-assisted selection as a component of conventional plant breeding. Plant Breed. Rev. 33, 145.

Gupta, D., Taylor, P.W.J., Inder, P., Phan, H.T.T., Ellwood, S.R., Mathur, P.N., et al., 2012a. Integration of EST-SSR markers of *Medicago truncatula* into intraspecific linkage map of lentil and identification of QTL conferring resistance to *Ascochyta* blight at seedling and pod stages. Mol. Breed. 30, 429–439.

Gupta, M., Verma, B., Kumar, N., Chahota, R.K., Rathour, R., Sharma, S.K., Bhatia, S., Sharma, T.R., 2012b. Construction of intersubspecific molecular genetic map of lentil based on ISSR, RAPD and SSR marker. J. Genet. 91, 279–287.

Gupta, P.K., Kulwal, P.L., Jaiswal, V., 2014. Association mapping in crop plants: opportunities and challenges. Adv. Genet. 85, 109–148.

Hamwieh, A., Udupa, S., Choumane, W., Sarker, A., Dreyer, F., Jung, C., et al., 2005. A genetic linkage map of *Lens* sp. based on microsatellite and AFLP markers and the localization of fusarium vascular wilt resistance. Theor. Appl. Genet. 110, 669–677.

Hamwieh, A., Udupa, S.M., Sarker, A., Jung, C., Baum, M., 2009. Development of new microsatellite markers and their application in the analysis of genetic diversity in lentils. Breed. Sci. 59, 77–86.

Havey, M.J., Muehlbauer, F.J., 1989. Linkages between restriction fragment length, isozyme and morphological markers in lentil. Theor. Appl. Genet. 77, 395–401.

Idrissi, O., Udupa, S.M., Houasli, C., De Keyser, E., Van Damme, P., De Riek, J., 2015. Genetic diversity analysis of Moroccan lentil (*Lens culinaris* Medik.) landraces using simple sequence repeat and amplified fragment length polymorphisms reveals functional adaptation towards agro-environmental origins. Plant Breed. 134, 322–332.

Kahraman, A., Kusmenoglu, I., Aydin, N., Aydogan, A., Erskine, W., Muehlbauer, F.J., 2004. QTL mapping of winter hardiness genes in lentil. Crop Sci. 44, 13–22.

Kahraman, A., Demirel, U., Ozden, M., Muehlbauer, F.J., 2010. Mapping of QTLs for leaf area and the association with winter hardiness in fall-sown lentil. Afr. J. Biotechnol. 9, 8515–8519.

Kaur, S., Cogan, N.O.I., Pembleton, L.W., Shinozuka, M., Savin, K.W., Materne, M., et al., 2011. Transcriptome sequencing of lentil based on second-generation technology permits large-scale unigene assembly and SSR marker discovery. BMC Genom. 12, 265. https://doi.org/10.1186/1471-2164-12-265.

Kaur, S., Cogan, N.I., Stephens, A., Noy, D., Butsch, M., Forster, J., et al., 2014. EST-SNP discovery and dense genetic mapping in lentil (*Lens culinaris* Medik.) enable candidate gene selection for boron tolerance. Theor. Appl. Genet. 127, 703–713.

Khazaei, H., Caron, C.T., Fedoruk, M., Diapari, M., Vandenberg, A., Coyne, C.J., McGee, R., Bett, K.E., 2016. Genetic diversity of cultivated lentil (*Lens culinaris* Medik.) and its relation to the world's agro-ecological zones. Front. Plant Sci. 7, 1093.

Kota, R., Rudd, S., Facius, A., Kolesov, G., Thiel, T., Zhang, H., et al., 2003. Snipping polymorphisms from large EST collections in barley (*Hordeum vulgare* L.). Mol. Gen. Genomics. 270, 24–33.

Kumar, S., Ali, M., 2006. GE interaction and its breeding implications in pulses. The Botanica 56, 31–36.

Kumar, J., Pratap, A., Solanki, R.K., Gupta, D.S., Goyal, A., Chaturvedi, S.K., et al., 2011. Advances in genomics resources for improving food legume crops. J. Agril. Sci. 150, 289–318.

Kumar, S., Hamweih, A., Manickavelu, A., Kumar, J., Sharma, T.R., Baum, M., 2014. Advances in lentil genomics. In: Gupta, S., Nadarajan, N., Gupta, D.S. (Eds.), Legumes in Omics Era. Springer Science+Business Media, New York, USA, pp. 111–130.

Kumar, J., Gupta, S., Gupta, P., Dubey, S., Tomar, R.S.S., Kumar, S., 2016. Breeding strategies to improve lentil for diverse agro-ecological environments. Indian J. Genet. 76 (4), 530–549.

Kumar, J., Gupta, D.S., Gupta, S., Dubey, S., Gupta, P., Kumar, S., 2017. Quantitative trait loci from identification to exploitation for crop improvement. Plant Cell Rep. 21, 1–27.

Kumar, J., Gupta, S., Biradar, R.S., Gupta, P., Dubey, S., Singh, N.P., 2018. Association of functional markers with flowering time in lentil. J. Appl. Genet. 59, 9. https://doi.org/10.1007/s13353-017-0419-0.

Mobini, S.H., Lulsdorf, T., Warkentin, D., Vanderberg, A., 2014. Plant growth regulators improve invitro flowering and rapid generation advancement in lentil and faba bean. In Vitro Cell Dev. Biol. Plant. https://doi.org/10.1007/s11627-014-9647-8.

Parisseaux, B., Bernardo, R., 2004. In silico mapping of quantitative trait loci in maize. Theor. Appl. Genet. 109, 508–514.

Perez de la Vega, M., Fratini, R.M., Muehlbauer, F.J., 2011. Lentil. In: Perez de la Vega, M., Torres, A.M., Cubero, J.I., Kole, C. (Eds.), Genetics, Genomic and Breeding of Cool Season Grain Legumes (Genetics, Genomics and Breeding in Crop Plants). CRC Press, pp. 98.

Phan, H.T., Ellwood, S.R., Hane, J.K., Ford, R., Materne, M., Oliver, R.P., 2007. Extensive macrosynteny between *Medicago truncatula* and *Lens culinaris* ssp. culinaris. Theor. Appl. Genet. 114, 549–558.

Reddy, M.R.K., Rathour, R., Kumar, N., Katoch, P., Sharma, T.R., 2009. Cross genera legume SSR markers for analysis of genetic diversity in *Lens* species. Plant Breed. 129, 514–518.

Rubeena, Ford, R., Taylor, P.W.J., 2003. Construction of an intraspecific linkage map of lentil (*Lens culinaris* ssp culinaris). Theor. Appl. Genet. 107, 910–916.

Rubeena, A., Taylor, P.W.J., Ades, P.K., Ford, R., 2006. QTL mapping of resistance in lentil (*Lens culinaris* ssp. *culinaris*) to ascochyta blight (*Ascochyta lentis*). Plant Breed. 125, 506–512.

Saha, G.C., Sarker, A., Chen, W., Vandemark, G.J., Muehlbauer, F.J., 2010a. Inheritance and linkage map positions of genes conferring resistance to stemphylium blight in lentil. Crop Sci. 50, 1831–1839.

Saha, G.C., Sarker, A., Chen, W.D., Vandemark, G.J., Muehlbauer, F.J., 2010b. Identification of markers associated with genes for rust resistance in *Lens culinaris* Medik. Euphytica 175, 261–265.

Saha, G.C., Sarker, A., Chen, W., Vandemark, G.J., Muehlbauer, F.J., 2013. Inheritance and linkage map positions of genes conferring agromorphological traits in *Lens culinaris* Medik. Int. J. Agron. 9. https://doi.org/10.1155/2013/618926.

Sato, S., Isobe, S., Tabata, S., 2010. Structural analyses of the genomes in legumes. Curr. Opin. Plant Biol. 13, 146–152.

Sever, T., Ates, D., Aldemir, S.B., Yagmur, B., Kaya, H.B., Temel, H.Y., et al., 2014. Identification QTLs controlling genes to Se uptake in lentil seeds. In: Proceedings of the Plant and Animal Genome XXII Conference, 2014.

Sharpe, A.G., Ramsay, L., Sanderson, L.A., Fedoruk, M.J., Clarke, W.E., Rong, L., et al., 2013. Ancient orphan crop joins modern era: gene-based SNP discovery and mapping in lentil. BMC Genom. 14, 1–13.

Solanki, R., Singh, S., Kumar, J., 2010. Molecular marker assisted testing of hybridity of F1 plants in lentil. Food Legum. 23, 21–24.

Steinhoff, J., Liu, W., Maurer, H.P., Würschum, T., Friedrich, C., Longin, H., Ranc, N., Reif, J.C., 2011. Multiple-line cross QTL mapping in European elite maize. Crop Sci. 51, 2505–2516.

Sudheesh, S., Rodda, M.S., Davidson, J., Javid, M., Stephens, A., Slater, A.T., Cogan, N.O., Forster, J.W., Kaur, S., 2016. SNP-based linkage mapping for validation of QTLs for resistance to ascochyta blight in lentil. Front. Plant Sci. 7. https://doi.org/10.3389/fpls.2016.01604.

Tadmor, Y., Zamir, D., Ladizinsky, G., 1987. Genetic mapping of an ancient translocation in the genus *Lens*. Theor. Appl. Genet. 73, 883–892.

Tahir, M., Muehlbauer, F.J., 1994. Gene-mapping in lentil with recombinant inbred lines. J. Hered. 85, 306–310.

Tanyolac, B., Ozatay, S., Kahraman, A., Muehlbauer, F.J., 2010. Linkage mapping of lentil (*Lens culinaris* L.) genome using recombinant inbred lines revealed by AFLP, ISSR, RAPD and some morphologic markers. J. Agric. Biotechnol. Sustain Dev. 2, 1–6.

Temel, H.Y., Gol, D., Kahriman, A., Tanyolac, M.B., 2014. Construction of linkage map through genotyping-by-sequencing in lentil. In: Proceedings of Plant and Animal Genome Conference. vol. XXII. pp. 358.

Thiel, T., Michalek, W., Varshney, R., Graner, A., 2003. Exploiting EST databases for the development and characterization of gene-derived SSR-markers in barley (*Hordeum vulgare* L.). Theor. Appl. Genet. 106, 411–422.

Tullu, A., Tar'an, B., Breitkreutz, C., Buchwaidt, L., Banniza, S., Warkentin, T.D., et al., 2006. A quantitative-trait locus for resistance to ascochyta blight *Ascochyta lentis* maps close to a gene for resistance to anthracnose *Colletotrichum truncatum* in lentil. Can. J. Plant Pathol. 28, 588–595.

Tullu, A., Tar'an, B., Warkentin, T., Vandenberg, A., 2008. Construction of an intraspecific linkage map and QTL analysis for earliness and plant height in lentil. Crop Sci. 48, 2254–2264.

Udupa, S.M., Robertson, L.D., Weigand, F., Baum, M., Kahl, G., 1999. Allelic variation at (TAA)n microsatellite loci in a world collection of chickpea (*Cicer arietinum* L.) germplasm. Mol Gen Genet 261, 354–363.

Varshney, R.K., Close, T.J., Singh, N.K., Hoisington, D.A., Cook, D.R., 2009. Orphan legume crops enter the genomics era!. Curr. Opin. Plant Biol. 12, 1–9.

Varshney, R.K., Mahendar, T., May, G.D., Jackson, S.A., 2010. Legume genomics and breeding. Plant Breed. Rev. 33, 257–304.

Verma, P., Goyal, R., Chahota, R.K., Sharma, T.R., Abdin, M.Z., Bhatia, S., 2015. Construction of a genetic linkage map and identification of QTLs for seed weight and seed size traits in lentil (*Lens culinaris* Medik.). PLoS One 10 (10), e0139666.

Weeden, N.F., Muehlbauer, F.J., Ladizinsky, G., 1992. Extensive conservation of linkage relationships between pea and lentil genetic maps. J. Hered. 83, 123–129.

Wurschum, T., 2012. Mapping QTL for agronomic traits in breeding populations. Theor. Appl. Genet. 25, 201–210.

Zamir, D., Ladizinsky, G., 1984. Genetics of allozyme variants and linkage groups in lentil. Euphytica 33, 329–336.

FURTHER READING

Kumar, S., Rajendran, K., Kumar, J., Hamwieh, A., Baum, M., 2015. Current knowledge in lentil genomics and its application for crop improvement. Front. Plant Sci. 6, 78.

Chapter 8

Genetic Transformation

Rakha Hari Sarker*, Subroto Kumar Das†, Kishwar Jahan Shethi*, M. Imdadul Hoque*

**Plant Breeding and Biotechnology Laboratory, Department of Botany, University of Dhaka, Dhaka, Bangladesh,* †*Department of Botany, University of Barisal, Bangladesh*

8.1. INTRODUCTION

Legumes have been cultivated as a major food crop since the early days of agriculture. These crops are particularly important as they all have high protein content useful for human and animal nutrition. Apart from this nutritional quality, the unique symbiotic ability of legumes in utilizing the atmospheric nitrogen for plant growth makes them preferable crops for sustainable agriculture (Oram and Agacoli, 1994, p. 31; ICARDA, 2000). Grain legumes belonging to the family Fabaceae are commonly known as pulses and play an important role in improving human diet. In many developing nations of the world, grain legumes have gained much importance in view of the acute shortage in production of animal proteins and due to the wide prevalence of protein malnutrition (Bressani, 1973). A number of grain legumes or pulses are being cultivated all over the world; these include lentil, chickpea, mung bean, black gram, garden pea, grass pea, cow pea, etc.

8.1.1 Lentil—A Popular and Important Pulse Crop

Lentils (*Lens culinaris* Medik.) are regarded as the most popular among the pulses and ranks first in terms of human consumption in different parts of the world. The importance of lentil lies in the fact that it is a major source of good quality protein in the common diet as its protein content can reach up to 24%–30% with 60% carbohydrate and 2.4%–4.2% mineral matter (Hulse, 1994, p. 242). It is a good source of cholesterol-lowering fiber and contains a significant amount of foliate, iron, and magnesium. Besides, lentils are also the main source of protein for livestock feed and inland fish production.

Lentils may have been one of the primitive agricultural crops grown more than 8500 years ago. It is also probably the oldest grain legume to be domesticated (Bahl et al., 1993) and has been the part of the human diet since Neolithic times. Lentils are supposed to have originated in southern Turkey

Lentils. https://doi.org/10.1016/B978-0-12-813522-8.00008-X

(Ladizinsky, 1979) and spread quickly to Greece, central and southern Europe, Egypt, Ethiopia, the Mediterranean, Afghanistan, the Indian Subcontinent, and China. It is presently cultivated in several other countries including Argentina, Colombia, Chile, Mexico, Peru, Syria, Canada, and the United States. Although it is a temperate crop, it is cultivated in the subtropics during winter months and at high altitudes in the tropics during colder months (Gowda and Kaul, 1982, p. 52). Areas with limited rainfall and dry growing seasons prove to be the most suitable for lentil production.

Lentils are a diploid herbaceous plant with $2n = 14$ chromosomes. Plants are typically short, with height ranging from 25 to 75 cm with soft hairy branches, slender and semi-erect with pinnately compound leaves having 4–7 pairs of leaflets. Flowers are papilionaceous and 4–8 mm long. The color of flower is white, lilac, or purple, depending on the genotype. Flowers are predominantly self-pollinated. Pods are flattened, 1–2 cm long, and they contain 1 or 2 seeds. Lentil seeds are classified in two major categories (Gowda and Kaul, 1982, p. 52): macrosperma and microsperma. Macrosperma have larger pods (15–20×7.5–10.5 mm) with large, flat seeds (6–8 mm dia). Cotyledons are yellow or orange. Flowers are large and white with light blue veins. Leaflets are large (15–27×4–10 mm) and oval. Plant height ranges from 25 to 75 cm. It is commonly available in the Mediterranean countries, Africa, and Asia Minor.

On the other hand, microsperma have small to medium pods (6–15×3.5–7 mm) which are convex. Seeds are flattened subglobose, small to medium (3–6 mm dia). Flower color is white to violet with varied patterns. Flowers are small (5–7 mm), with elongated peduncles. Leaflets are small (8–15×2–5 mm) elongated, linear, or lanceolate. Height of plants varies from 15 to 35 cm. They are found in the southwestern, western, and eastern areas of Asia.

It is evident that, due to its nutritional properties, cooking quality, and easy digestibility, the demand for this crop has been steadily increasing in many countries. On a global scale, lentil consumption is rising among the cool season pulses, and it is by far the fastest growing crop. Therefore, it is expected that by 2030, world lentil consumption will be doubled.

8.1.2 Constraints to Lentil Production—Biotic and Abiotic Stress Factors

Despite the fact that the lentil is one of the mankind's oldest cultivated crop plants, its agronomic performance resembles that of a wild plant, rather than a cultivated plant. Production of lentils is usually characterized by low yield potential in most of the lentil-growing countries. Several factors are believed to be responsible for the lower production of this important pulse crop, including susceptibility to disease and pests, massive flower drop, postharvest loss, and management problems (Erskine, 1997; Erskine and Sarker, 1997). Lentil diseases at various stages of growth are caused by fungi, bacteria, viruses, insect pests, and nematodes. Among these constraints, fungal diseases cause the maximum damage. In the Indian subcontinent, the major fungal diseases are *Stemphylium* blight, caused by *Stemphylium sarciniformis*; rust, caused by *Uromyces fabae*; root rot, caused by *Rhizactonia solani* and wilt, caused by *Fusarium oxysporium.* Ascochyta blight is a serious threat for the lentil growers of India, Canada, and Australia. These diseases cause variable degrees of yield loss ranging from zero to 100%, depending on the severity of infection (Rahman and Bakr, 1998, p. 27). The major insect pests causing damage to lentil are weevil, pod borer, thrips, grain beetle, aphids, bruchids, etc. (Gowda and Kaul, 1982, p. 73). Healthy lentil plants of microsperma type growing in the field are presented in Fig. 8.1, while a *Stemphylium* blight-affected lentil plant is presented in Fig. 8.2. Photographs of healthy mature seeds of the

FIG. 8.1 Healthy lentil plants growing in the field.

FIG. 8.2 Lentil plants in the field with severe symptoms of *Stemphylium* blight.

FIG. 8.3 Mature dry seeds of a microsperma variety of lentil.

FIG. 8.4 Lentil seeds heavily infested with storage bruchids.

microsperma type of lentil and the same seeds severely infested with storage bruchids are depicted in Figs. 8.3 and 8.4, respectively.

Apart from these biotic stress factors, several abiotic stresses such as cold, drought, heat, and salinity affect lentil production worldwide (Saxena, 1993; Monti et al., 1994). Salinity problems with lentil are not widespread, but can be acute in certain regions of the Indian subcontinent, the Nile Delta of Egypt, and in some areas of Turkey. Canada also has some difficulty in lentil production due to high salinity areas present in Saskatchewan (Katerji et al., 2003; Muehlbauer and McPhee, 2002). Thus, the major constraints for lentil breeding and production are poor yield stability and high susceptibility to diseases and pests.

8.1.3 Improvement of Lentils Through Conventional Breeding

The yielding ability of existing lentil varieties is inadequate as several biotic and abiotic stresses have imposed limitations on the development of high-yielding varieties of lentil worldwide. Therefore, to obtain desired yield performance, improvement of this crop is essential. There is a need to increase productivity, and to enhance the nutritional value and other desired agronomic qualities of this crop.

Conventional breeding methods including hybridization techniques, and selections are being carried out to develop improved varieties of many crops with higher yield, better quality, wide adaptability, and disease resistance (Zohary, 1986; Shimelis and Laing, 2012). In the past, several attempts have been undertaken to develop disease-resistant, high-yielding varieties of lentil through distant hybridization and mutation breeding. However, in lentils, the progress of improvement through conventional breeding is hampered due to a lack of adequate genetic variability which is caused by a predominantly high degree of self-pollination and absence of resistance gene/s of interest in the existing germplasm of lentil (Khazaei et al., 2016). Moreover, crossing in lentils is tedious because of the small size of their flowers, and hybridization between microsperma and macrosperma is hampered due to asynchronous flowering, although development of viable hybrids between different groups of lentil through embryo rescue has been reported (Ladizinsky et al., 1985).

Several other important traits, such as biomass yield, pod shedding, nitrogen fixation, resistance to pea weevil, aphids, and parasitic weeds, namely, *Orobanche* and *Striga*, cannot currently be addressed through conventional breeding because of insufficient genetic variation. Plant breeders have been trying to identify resistant/tolerant germplasm and to generate genetic map positions of the resistant genes; moreover, molecular breeding of lentils for disease-resistant genes using marker-assisted selection is progressing gradually (Muehlbauer and McPhee, 2002), which may contribute in the future improvement of this crop.

8.1.4 Application of Biotechnology for Improvement of Lentil

Because conventional breeding has several limitations, it is imperative to look for some other methods to induce genetic variability in lentil. Besides conventional breeding techniques, there are many ways of creating variability, including induction of somaclonal variation (Larkin and Scowcroft, 1981), somatic hybridization, and plant genetic engineering. But the techniques of somaclonal variation and somatic hybridization were found to have limited application in many crop species, including lentils. Under these circumstances, modern techniques of biotechnology, commonly known as "plant genetic transformation" can be applied for the improvement of this crop. Genetic transformation is an emerging technology that can supplement traditional crop improvement procedures, and this approach can be used in introducing desired traits, which is not possible through conventional breeding alone (Gardner, 1993). Thus, genetic transformation enhances and broadens the genetic variability in several cases where the natural variability within a species is not sufficient. This technology is regarded as a pre-breeding method that provides a solution to certain constraints that limit crop production and quality. Therefore, it might be possible that the genetic transformation method, combined with traditional breeding techniques, could aid in improving both the quality and yield of lentils.

The biotechnological approaches for the improvement of grain legumes can mainly be focused on *Agrobacterium*-mediated genetic transformation, biolistics for gene delivery, electroporation, and/or polyethylene glycol (PEG) treatments. Success of transformation in leguminous crops is limited because of their recalcitrant nature under in-vitro conditions (Nisbet and Webb, 1990). Nevertheless, *Agrobacterium*-mediated transformation has been demonstrated in a number of leguminous crops such as soybean (Hinchee et al., 1988; Jacobson, 1992), pea (Schroeder et al., 1993), subterranean clover (Khan et al., 1994), and chickpea (Kar et al., 1996; Krishnamurthy et al., 2000). Under these circumstances, there is considerable scope for embracing biotechnological approaches for the improvement of lentils.

8.1.4.1 Status of In-Vitro Regeneration

The major prerequisite for the application of biotechnology, particularly for the exploration of plant genetic transformation, is the availability of an effective in-vitro regeneration system for the particular plant of interest. In-vitro regeneration and transformation procedures of lentil are not well developed compared to the success achieved in other grain legumes from Europe and North America. In-vitro culture of lentils has proven to be difficult because of its recalcitrant nature. During the last several years, techniques of plant regeneration in lentils have progressively improved.

The first partial success of in-vitro plant regeneration in lentils was obtained from meristem tip culture (Bajaj and Dhanju, 1979). Later, a protocol for regeneration of lentils from the hypocotyl and epicotyl-derived callus cells was reported (Williams and McHugen, 1986).

Further formation of complete plantlet using callus tissue induced from embryonic axes was achieved (Saxena and King, 1987). Moreover, formation of multiple shoots from shoot tip, the first node, and the first pair of leaves in culture media supplemented with 6-benzyladenine (BA) and naphthalene acetic acid (NAA) was obtained (Polanco et al., 1988).

Using seed culture multiple shoot formation from cotyledonary nodes on a BA-fortified medium was also reported (Malik and Rashid, 1989). However, lentil plants have been reported to be regenerated directly from nodal segments and shoot tips as well as from callus cells (Singh and Raghuvanshi, 1989). It was also noticed that nodal segments and shoot tips produced multiple shoots on a medium supplemented with kinetin (Kn) without the intervention of callus. MS (Murashige and Skoog, 1962) medium supplemented with thiadiazuron (TDZ) was found to be effective in producing multiple shoots by culturing lentil seeds (Malik and Saxena, 1992), while multiple shoot formation from cotyledonary nodes using BA was also reported (Warkentin and McHughen, 1993).

In case of microsperma varieties of lentils, reports are available about in-vitro regeneration of shoots from the explants of cotyledonary nodes and decapitated zygotic embryo on a medium supplemented with benzyl amino purine (BAP), giberellic acid (GA_3), and Kn (Sarker et al., 2003b). These studies indicate that cytokinins induce multiple shoot formation in different types of explants from lentils. Upon exploration of high regeneration capacity of lentil explants, a high concentration of cytokinin is used in tissue culture media. However, during the rooting procedure, it was understood that the impact of cytokinin usage was depressive to root regeneration. This issue was adequately covered in a study that showed the effect of BAP on root formation from lentil shoots regenerated on BAP-containing media (Polanco and Ruiz, 1997).

Studies have also indicated that rooting in the microsperma type of lentil is especially difficult and appears to be a major constraint in obtaining fully regenerated plants through in-vitro techniques. A small number of roots from the regenerated shoots of the microsperma type of lentil produced from BAP-containing media when a high concentration of auxin was used. But these induced roots mostly did not come from the base of the cut end of the shoots, and after some days, they behaved abnormally, with callus deposited at the tip of the roots, and afterwards, the shoots failed to survive (Khanam et al., 1995; Sarker et al., 2003b).

This dilemma of rooting problems versus the demand for a higher number of putative transgenic shoots from transformation studies of lentil were partially resolved (Gulati et al., 2001) when lentil (macrosperma type) shoots regenerated on BAP-containing media were micrografted onto 5–6 day-old rootstocks with 96% success efficiency. The success of micrografting was noted to be independent of the nature and concentration of growth regulator used in shoot initiation medium and the time period for induction of shoots. The regenerated lentil shoots from cotyledonary node explants were reported to be micrografted with high efficiency (Khawar and Ozcan, 2002). Unfortunately, this approach of micrografting was fraught with difficulties, especially in the case of microsperma types of lentil.

Under these circumstances, in-vitro flowering and fully developed seed formation have been considered as a viable alternative to obtain plant progenies. It is believed that in-vitro flowering and pod production may significantly contribute toward the genetic improvement of lentils through a biotechnological approach.

8.1.4.2 Status of Genetic Transformation

To date, a number of reports are available on attempts to transform lentil tissue. The tumor-producing ability of four diverse strains of *Agrobacterium tumefaciens* in shoot and root apices of lentil under field conditions was the first report about the attempts of *Agrobacterium*-mediated transformation of lentils

(Warkentin and McHughen, 1991). This study demonstrated that the tumor formation and opine production ability were indicative of plant cell transformation. Lentil shoot apex, epicotyl, and root explants were capable of expressing an intron-containing *GUS* reporter gene after inoculation with the disarmed *Agrobacteriun* strain. Expression occurred at all wound sites on these explants except at the end of root explants proximal to the cotyledonary node, and the axils of the cotyledonary petioles were not amenable to transformation (Warkentin and McHughen, 1992).

Co-electroporation of plasmid DNA in lentil root protoplasts was also reported (Maccarrone et al., 1995). Moreover, using intact nodal meristems *in planta* transformation was demonstrated (Chowrira et al., 1995). In this study, *GUS* reporter gene was used to monitor the nature of the transformed tissue, and this experiment could recover transgenes from the offspring of electroporated individuals. This investigation also reported to have recovered chimeric branches from the R1 seeds developed through electroporation of *GUS* reporter gene in the intact plant tissue of the lentil (Chowrira et al., 1996).

Expression of the *GUS* gene has also been observed after inoculation of a longitudinally sliced embryogenic axis of lentils with different *Agrobacteriun* strains (Lurquin et al., 1998). Further transformation experiments were carried out using particle gun bombardment subjecting cotyledonary nodes of lentils where 50% GUS expression was obtained in the bombarded explants (Öktem et al., 1999). However, none of these studies demonstrated the successful development of a complete plantlet expressing foreign gene/s in lentil.

The technique of vacuum infiltration has been reported to enhance transformation frequency in *Agrobacterium*-mediated gene transfer in some legumes (Trieu et al., 2000). Accordingly, the first report of a vacuum infiltration *Agrobagterium*-mediated transient expression system on lentil cotyledonary nodes was reported (Mahmoudian et al., 2002; Celikkol et al., 2009). These results showed that infiltration increases the efficiency of transformation, but a low number of regenerated shoots exhibit GUS expression.

Apart from these studies, *Agrobacterium*-mediated transformation system was developed for two microsperma varieties of lentil (Sarker et al., 2003a), where the highest percentage of GUS positive regions was found in epicotyl explants followed by decapitated embryo explants. Stable integration of *GUS* and *npt*II genes was confirmed by polymerase chain reaction analysis. *Agrobagterium*-mediated transformation was carried in lentils using different tissues including peeled cotyledonary nodes, cotyledonary petioles, shoot tips, and roots as explants on various media via indirect organogenesis. In this investigation, shoot tips exhibited the highest percentage of GUS expression, and root formation was only achieved in a media containing auxin supplements such as indole-3-acetic acid (IAA) or NAA (Bayrac, 2004).

The tumor and root formation ability of several tissues from different lentil cultivars after inoculation with *Agrobacterium rhizogens* and *Agrobacterium tumefaciens* strains was compared where the frequency of tumor formation from cotyledonary node explants was reported to be higher compared to shoot meristems (Dogan et al., 2005).

Later, optimization of lentil transformation mechanism was carried out using cotyledonary node tissues with the combinations of several treatments in three strains of *Agrobacteriun tumefaciens*. In this study, the fertile transgenic plants were recovered only through grafting of transgenic shoots on rootstocks. Moreover, the transgene insertion and expression were confirmed through PCR, RT-PCR, and Southern hybridization, and the transgenic progenies were found to segregate according to Mendelian fashion (Akcay et al., 2009).

In order to enhance drought and salinity tolerance, the researchers tried to introduce the *DREB1A* gene in lentils. This *DREB1A* gene was driven by the rd29A promoter into lentil decapitated embryo explants, followed by shoot regeneration from the apical meristems of cotyledonary buds via direct organogenesis. Subsequently basta-resistant putative transgenic explants were micrografted onto nontransformed rootstocks to establish transgenic plants. Transgenic insertion and inheritance to the progeny were evaluated through PCR and Southern blot analysis. Expression of *DREB1A* gene in transgenic plants introduced by salt stress was confirmed through RT-PCR (Khatib et al., 2011).

Moreover, several investigations indicated that the choice of suitable explants was very important factor for achieving efficient transformation in lentil. It has been pointed out that the explants which produce very good response in regeneration system, show low response in transformation studies. In a study, it was observed that above 90% of the cotyledonary node explants regenerated shoots, but no recovery of transformed shoots was noticed in *Agrobacterium*-mediated genetic transformation studies, whereas decapitated embryo showed comparatively less response for in-vitro regeneration but produced very high response in obtaining transformation efficiency (Sarker et al., 2003a).

Besides these, another most important difficulty in legumes is the integration of foreign genes through genetic transformation. Various methodologies that are utilized for the introduction of foreign DNA into leguminous crops also illustrate that no single technique is optimal for the transformation of all legumes because of the species' specificity toward various methods (Atkins and Smith, 1997, p. 296). The transformation rate of various explants was always found to be low, and in most cases, about 1% of transformation efficiency was recorded.

8.1.4.3 Strategies for the Development of Fungal Disease Resistance

Fungal diseases are the most important factors contributing toward yield loss in lentil. These diseases can partly be controlled by widespread application of fungicide. The economic cost and negative environmental impact associated with this application have led to the recent search for alternative strategies like plant genetic transformation for the development of fungal disease resistance in lentil cultivars. Antifungal transgenic approaches are primarily associated with the transfer of *chitinase* and other antifungal protein (*AFP*) genes in achieving fungal disease resistance in lentils.

Among various fungal-resistant genes, *chitinase* genes are potentially the most promising, as the enzymes degrade the substance chitin found in fungal cell wall, and moreover, they showed activity against a wide range of fungus (Lorito et al., 1998). Hence, genetic engineering of plants with *chitinase* gene is attractive for fungal disease control mechanism. As chitinase are one of the most important protein families showing antifungal activity in vitro, several reviews and research articles (Neuhaus et al., 1991; Nagasaki et al., 1997; Fukamizo et al., 2001; Van den Bergh et al., 2004) have stressed the advantages of using the *chitinase* gene in plants to enhance resistance to fungal pathogens (Gao et al., 2000; Islam, 2006; Nirala et al., 2010). On the other hand, antifungal proteins from different sources showed in-vitro activity by inhibiting hyphal and fungal growth as reported by many scientists (Asao et al., 1997; Bolar et al., 2000; Rajasekaran et al., 2000).

The most important prerequisite for plant genetic transformation is the availability of a suitable and reproducible transformation protocol. Sufficient studies have not been carried out for the improvement of lentils using modern techniques of plant genetic transformation. Recently limited progress has been made for both in-vitro regeneration and transformation of microsperma lentil varieties cultivated in Bangladesh (Sarker et al., 2003a, 2003b; Hassan et al., 2007). The overall progress in genetic transformation of lentils is mostly hampered by the lack of a suitable transformation compatible, as well as a genotype-independent in-vitro regeneration system (Hoque et al., 2004, p.175), and more precisely, *Agrobacterium*-mediated genetic transformation in lentils is hampered by their recalcitrant nature, particularly pertaining to the development of an effective in-vitro root induction system.

In this chapter, we report on the progress made so far toward the development of an efficient in-vitro regeneration system, leading to an effective protocol for *Agrobacterium*-mediated genetic transformation in microsperma varieties of lentils, as well as for the integration of fungal disease resistance genes in lentil through *Agrobacterium*-mediated genetic transformation.

8.2. PROTOCOLS FOR IN-VITRO REGENERATION OF PLANTLETS

Genetic transformation compatible in-vitro regeneration system is required for the development of an effective transformation protocol for lentils. The transformation procedure for recalcitrant lentil is not well established particularly due to the lack of a complete protocol for in-vitro regeneration. Moreover, only limited information is available for in-vitro regeneration of the microsperma type of lentils. In this context, special emphasis was given to the development of an efficient, reproducible, genotype-neutral, in-vitro regeneration system for the microsperma type of lentils, so that this protocol can be applied dependably to obtain stable transformants of lentils.

8.2.1 Plant Materials Used for In-Vitro Regeneration

Four microsperma varieties of lentils (*Lens culinaris* Medik.) namely, BARI masur-1 (BM-1), BARI masur-4 (BM-4), BARI masur-5 (BM-5), and BARI masur-6 (BM-6), typically cultivated in Bangladesh were used for in-vitro regeneration. Seeds of these four varieties of lentils were collected from the Bangladesh Agricultural Research Institute (BARI), Joydebpur, Gazipur, and maintained in the Plant Breeding and Biotechnology Laboratory of the Department of Botany, University of Dhaka, Bangladesh.

8.2.2 Preparation of Explants for In-Vitro Regeneration

Three different zygotic embryo derived explants, namely cotyledonary node (CN), decapitated embryo (DE) and cotyledon-attached decapitated embryo (CADE), were used for in-vitro regeneration.

CN explants were collected from aseptically germinated seedlings. For the preparation of CN explants, lentil seeds were first washed in 70% ethanol for 1 min, and then surface sterilized with 0.1% $HgCl_2$ (*w/v*) solution for 15 min. Seeds were then rinsed three or four times with sterilized distilled water. The surface-sterilized seeds were then cultured on 0.3% (*w/v*) water-agar medium for germination. Finally, CN explants were prepared by removing the root and shoot tip as well as half of both cotyledons from three-day-old seedlings. The explants of DE and CADE were prepared from seeds soaked overnight by splitting them open. In case of DE, root and shoot tips from each embryo, along with cotyledons, were removed, whereas CADE has a decapitated embryo, but with cotyledon attached.

8.2.3 Development of In-Vitro Shoots

Explants of cotyledonary node (CN), decapitated embryo (DE), and cotyledon-attached decapitated embryos (CADE) were cultured on MS (Murashige and Skoog, 1962) medium supplemented with 2.22 μM 6-benzylaminopurine (BAP), 2.32 μM kinetin (Kn), 0.29 μM gibberellic acid (GA_3), 30.35 μM tyrosine, and 3% sucrose (*w/v*) in a 250 mL Erlenmeyer flask containing 60 mL of medium for multiple shoot regeneration (Sarker et al., 2003b; Das et al., 2012). In all cases, the pH of the medium was adjusted to 5.8 prior to adding 0.8% agar (*w/v*), and autoclaved at 121°C for 20 min. The culture vessels were incubated in a growth room with a 16/8 h light/dark cycle at 25 ± 2°C.

8.2.4 Development of In-Vitro Roots From Excised Regenerated Shoots

Development of roots from in-vitro raised shoots is an essential step to obtain complete plantlets. For induction of root growth, 2–4 cm long regenerated healthy shoots were excised and transferred to test tubes (25×150 mm), each containing 10–15 mL of MS or half-strength MS medium supplemented with indole-3-butyric acid (IBA) ranging from 49.2 to 123.0 μM. The test tubes were also incubated in the growth room under a 16/8 h light/dark cycle at 25 ± 2°C for induction of roots.

8.3. PROTOCOL FOR IN-VITRO FLOWER AND POD FORMATION

Improvement of lentil through biotechnological methods has proven to be difficult due to its recalcitrant nature, particularly pertaining to the development of an effective in-vitro root induction system. Previous studies have indicated that rooting in the microsperma type of lentil is especially troublesome and appears to be a major constraint in obtaining fully regenerated plants through in-vitro techniques (Khanam et al., 1995; Sarker et al., 2003b). Therefore, in-vitro flowering and fully developed seed formation have been considered as a viable alternative in obtaining required plant progenies. It is believed that in-vitro flowering and pod production may significantly contribute toward the genetic improvement of lentil through a biotechnological approach.

To obtain in-vitro flower and pod formation, cotyledonary nodes (CN) and cotyledon-attached decapitated embryo (CADE) explants were cultured for the development of shoots only, because DE explants showed the least response toward shoot regeneration. After 15 to 20 days of culture, when the regenerated shoots attained a length of 2–3 cm, individual elongated shoots were separated and cultured separately in test tubes (25 × 150 mm) containing 10 mL of auxin-rich MS medium for in-vitro flower formation. Auxins used for the induction of flowers were IBA or indole-3-acetic acid (IAA), and 1-naphthaleneacetic acid (NAA). Cultures were maintained in the growth room until flower and pod formation (Sarker et al., 2012).

8.3.1 Pollen Viability and Pollen Tube Growth Following the Self-Pollination

Monitoring the post pollination events following the self-pollination in case of in-vitro grown flowers is required to understand the nature of pollination leading to the formation of pods and seeds under in-vitro conditions. To examine the viability of pollen grains, in-vitro-derived individual flowers were excised, sepals and petals were removed, and the anthers were placed on a clean slide to squash them in a solution containing 0.01% fluorescin diacetate (FDA) and 20% sucrose, and incubating them for 30 min (Heslop-Harrison and Heslop-Harrison, 1970). After that, all the debris were removed and the pollen grains were observed under a fluorescence microscope.

To study the growth of pollen tubes in situ, the self-pollinated pistils from in-vitro-grown flowers were collected, fixed, and softened in a mixture of acetic acid:alcohol (1:3, *v/v*) with 1 M NaOH at 60°C for 10 min, and thereafter stained with 0.1% (*w/v*) decolorized aniline blue solution. (The decolorized aniline blue solution was prepared by dissolving 0.1 g of aniline blue in 100 mL of 0.1 M tripotassium orthophosphate [Sigma-Aldrich Chemical Co., United States] followed by storage of the solution in an amber glass bottle for at least 1 week at room temperature until the solution became pale green.) After staining, pistils were mounted in one drop of 50% (*v*/*v*) aqueous glycerol solution prior to microscopic examination. Stained pistils and pollen grains were observed under a fluorescence microscope fitted with an incident UV illumination system (Nikon, Microphot; excitation 450–490 nm) (Kho and Baer, 1968).

8.3.2 Propagation of In-Vitro Grown Seeds

Mature seeds developed from in-vitro-grown flowers were harvested and dried under sunlight approximately for 2–3 days. Air-dried seeds were surface disinfected and placed on water-agar medium for germination, as described previously. After 2–3 weeks following germination, seedlings were transplanted into 10 × 6 cm plastic pots containing sterilized peat moss. The roots were gently washed to free them from agar using distilled water. Pots were covered with transparent, perforated polythene bags. The inner side of these bags was moistened with water to prevent desiccation. Plants were acclimatized in the culture room at 25 ± 2°C for 1–2 weeks, and then further maintained in a net house for flower and pod development.

8.4. PROTOCOL FOR *AGROBACTERIUM*-MEDIATED GENETIC TRANSFORMATION

Genetically engineered *Agrobacterium tumefaciens* strains having three different plasmid vectors containing three different gene constructs were used in genetic transformation experiments. Although four varieties of lentils were used for in-vitro regeneration, but the best responsive three varieties, namely, BM-4, BM-5, and BM-6, were employed for *Agrobacterium*-mediated genetic transformation using the three different types of explants, namely, cotyledonary node (CN), decapitated embryo (DE), and cotyledon-attached decapitated embryo (CADE). The properties of plasmid vectors and the gene constructs carried by them have been described in the next section.

8.4.1 Gene Construct I (pBI121*GUS npt*II)

Agrobacterium tumefaciens strain LBA4404 contained plasmid pBI121 (binary vector). This binary vector had two genes within the T-DNA region of the gene construct (Fig. 8.5A). The binary vector pBI121 has the background of pBIN19. It contained a scorable reporter gene *GUS* (β-glucurunidase) driven by a CaMV35S promoter and NOS terminator and a selectable marker gene *npt*II fused between the NOS promoter and the NOS terminator encoding for the enzyme neomycin phosphotransferase conferring kanamycin resistance (Herrea-Esterella et al., 1983).

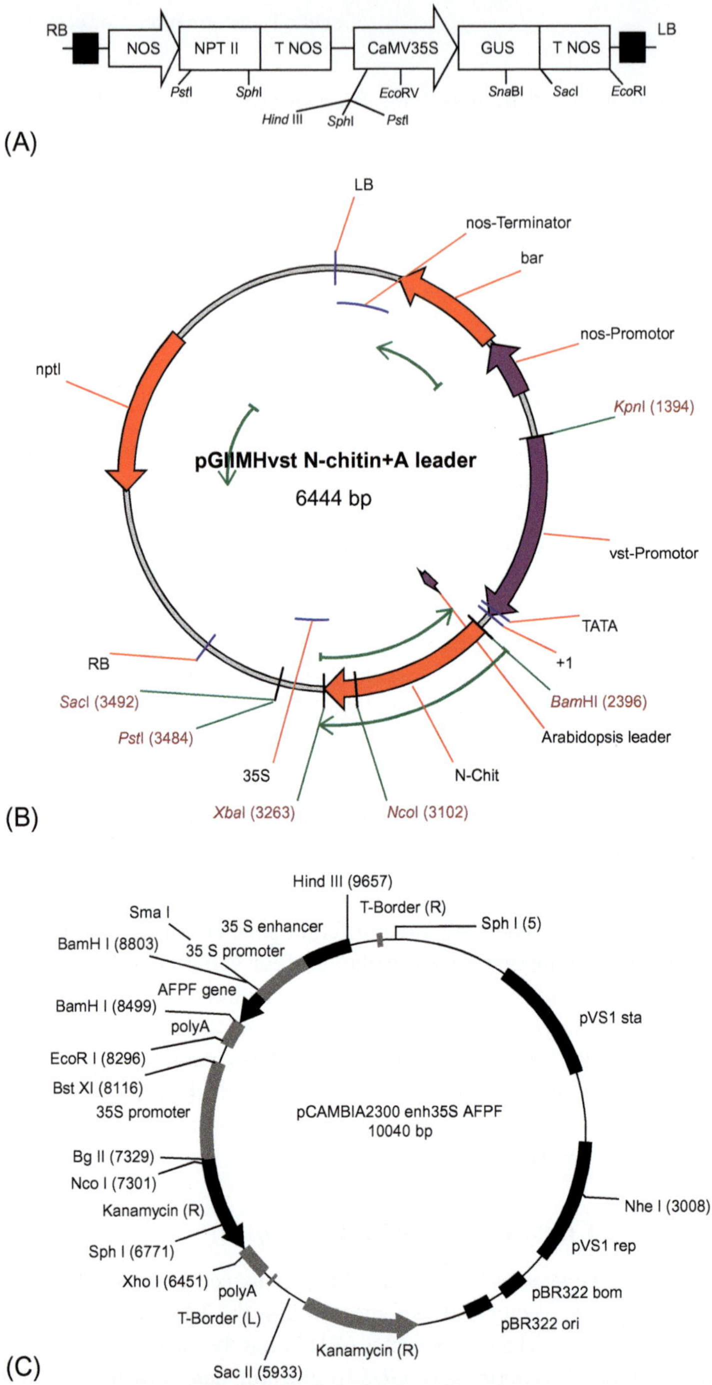

FIG. 8.5 Diagrammatic representation of *Agrobacterium tumefaciens* strains containing: (A) Gene construct I: *Agrobacterium tumefaciens* strain LBA4404 containing plasmid pBI121 showing a part of T-DNA between left and right border; (B) gene construct II: *Agrobacterium tumefaciens* strain EHA105 containing plasmid pGreen0229 containing pGIIMHvstN-chitin; (C) gene construct III: *Agrobacterium tumefaciens* strain LBA4404 containing plasmid pCAMBIA2300enh35SAFP.

The reporter *GUS* gene was used to assess the efficiency of transformation. The *GUS* gene has an intron in the coding sequence which can express GUS activity in the plant cell but not in the cells of *Agrobacterium*. In this case, transformed tissues as well as transformed plantlets were selected by using kanamycin. This gene construct was specifically designed for the development of genetic transformation protocol in lentil.

8.4.2 Gene Construct II (pGIIMHvst N-chitin)

The hypervirulent *Agrobacterium tumefaciens* strain EHA105 (Hood et al., 1993) harbors the pGreenII 0229 derivative binary plasmid containing a *bar* gene and the pSoup helper plasmid, (Hellens et al., 2000). This binary vector contains following genes within the right border (RB) and left border (LB) region of the construct (Fig. 8.5B).

Selectable marker *bar* gene encoding phosphinothricin-acetyl transferase is driven by NOS promoter and NOS terminator and *chitinase* gene isolated from *Streptomyces olivaceoviridis* (ATCC 11238). Chimeric constructs were developed by fusing the *Arabidopsis thaliana* leader peptide (NCBI, AY081519) to the *chitinase* gene behind the stilbene synthase promoter (pGIIvstN-Chitin) from grape (Wiese et al., 1994). This gene construct was kindly provided by Professor H. J. Jacobsen, Department of Molecular Genetics, Leibniz University of Hannover, Germany.

8.4.3 Gene Construct III (pCAMBIA2300enh35SAFP)

Agrobacterium tumefaciens strain LBA4404 contains plasmid pCAMBIA2300enh35SAFP. This binary vector contains *npt*II and *AFP* genes within the right border (RB) and left border (LB) region of the construct (Fig. 8.5C). This is considered as an antifungal protein gene construct and was kindly provided by Dr. V. S. Reddy, International Centre for Genetic Engineering and Biotechnology (ICGEB), New Delhi, India. In this case, transformed plantlets were selected by using kanamycin, as this gene construct contained a kanamycin-resistant gene.

8.4.4 Steps of Transformation and Selection of Transformants

To develop a healthy culture of *Agrobacterium*, 50 mL of liquid YMB (Hooykaas, 1988) containing 50 mg/L kanamycin was inoculated with *Agrobacterium* from a fresh bacterial plate and grown at 180 rpm on a rotary shaker at 28°C for 16 h. The overnight grown *Agrobacterium* culture was centrifuged for 10 min at 5000 rpm and the pellet was resuspended in liquid MS medium (pH 5.8) to make the required *Agrobacterium* suspension. This *Agrobacterium* suspension was used for infection and incubation of explants. Prior to this, the "Optical Density" (OD) of the bacterial suspension was determined at 600 nm with the help of a spectrophotometer (Shimadzu, Japan). In obtaining sufficient infection, the cut explants were dipped in bacterial suspension for different incubation periods before transferring them to the co-cultivation medium (MS + 2.22 μM BAP + 2.32 μM Kn + 0.29 μM GA_3 + 30.35 μM tyrosine).

Following infection and incubation, the explants were soaked in sterile Whatman filter papers for a short period to remove excess bacterial suspension. All the explants were maintained in coculture medium for 2–4 days in dark condition in the growth room at 25 ± 2°C.

Following coculture, the explants were washed with distilled water three or four times until no opaque suspension was seen, then washed for 15 min with distilled water containing 300 mg/L ticarcillin (Duchefa, The Netherlands). Then, explants were dried with a sterile Whatman filter paper and transferred to regeneration medium with 100 mg/L ticarcillin. After 7–10 days, the regenerated shoots were then subcultured in selection medium containing 50 mg/L kanamycin and either 100 mg/L ticarcillin or 50 mg/L ticarcillin. Cultures were subcultured regularly at an interval of 12–15 days, and the concentration of kanamycin was gradually increased up to 200 mg/L on the selection medium. Shoots survived on selection medium were sub-cultured on ½ strength of MS medium containing 98.420 μM IBA and 2.69 μM NAA with 50 mg/L kanamycin subjected to in-vitro flowering and seed formation when required. Kanamycin was used for the selection of transformants derived following transformation experiments using gene constructs I and III. On the other hand, for construct II, PPT (phosphinothricin) was used as the selective agent because this gene construct contained *bar* gene, which was resistant to PPT. To find out the appropriate concentration of selective agent, different concentrations (0.5–2.0) of PPT were applied.

8.4.5 GUS Histochemical Assay

The transformation ability of the explants was monitored through GUS histochemical assay (Jefferson et al., 1987; Gould and Smith, 1999) and gave an insoluble indigo blue color at the site of GUS enzyme activity. Thus, it allowed transformed tissues to be screened histochemically. Tissues and shoots under selection pressure of around 2 months were monitored for stable GUS expression.

8.4.6 Molecular Analysis in Detecting Expression of Transgenes

For molecular analysis, the total genomic DNA was extracted from fresh leaves of putative transformants and control plants by CTAB method (Doyle and Doyle, 1990). The presence of the *GUS*, *npt*II, and *AFP* genes in the lentil genomic DNA were analyzed by polymerase chain reaction. For the detection of the *npt*II coding sequence, DNA was subjected to PCR using the following primers and conditions: For the *npt*II gene coding sequence, forward 5′-TGA

TTG AAC AAG ATG GAT TG-3′ and reverse 5′-CAT TTT CCA CCA TGA TAT TC-3′ were utilized. For the *GUS* gene, the primers were forward 5′-CCT GTA GAA ACC CCA ACC CG-3′ and reverse 5′-TGG CTG TGA CGC ACA GTT CA-3′ (MGW-Biotech, AG, and Germany). Primers for *npt*II and *GUS* marker genes amplified DNA products of 700 and 500 bp, respectively.

Sequences for the primers amplifying *chitinase* gene were forward 5′-GGT GAC ATC GTC CGC TAC AC-3′ and for reverse, 5′-GGT GTT CCA G-TA CCA CAG CG-3′. Product size of *chitinase* gene is 555 bp.

Sequences for the primers amplifying *AFP* gene were forward 5′-CGC GGA TCC ATG GCG AGG TGT GAG AAT TTG GCT-3′ and for reverse, 5′-GAG TCC TCG AGA CAA TTT TTG GTG CAC CAA CAA C-3′. Primers for *AFP* gene amplified 190 bp products.

All primers were used at a concentration of 100 pmol/μL. The plasmid pBI121 isolated from *Agrobacterium tumefaciens* was used as the positive control. PCR reaction mix of 25 μL contained 2.5 μL of 10×PCR buffer with 15 mM $MgCl_2$ (Gene Craft, Germany), 1 μL of 5 mM of the dNTP mix, 1 μL of Red Taq polymerase (Natutech, Germany), 1 μL of each of the respective primers, and 1 μL (50–80 ng/μL) of the sample DNA and 17.5 μL ultra-pure water. For PCR amplification of the *GUS* gene, DNA was denatured at 94°C for 3 min and then amplified in 30 cycles using 94°C for 1 min, 64°C for 1 min (annealing), and 72°C for 1 min followed by 5 min at 72°C. For the *npt*II gene, the cycling conditions were 3 min at 94° denaturation and 30 amplification cycles using 94°C for 1 min, 55°C for 1 min (annealing), and 72°C for 1 min followed by 5 min at 72°C. The amplified DNA was run on 1.0% agarose gel and stained with ethidium bromide (0.05 μg/mL) and photographed.

Total RNA was isolated by using TRIZOL reagent following the manufacturer protocol and first-stand cDNAs were synthesized by using the templates RNAs following the procedure of affinity script QPCR cDNA synthesis kit (Agilent Technologies, Integrated Sciences). Synthesized cDNA strands were directly used in the reverse transcription PCR (RT-PCR) using gene specific primers both for the amplification of *npt*II and *AFP* gene.

Southern and Northern blot analysis was performed by using the standard protocol (Sambrook et al., 1989). For the preparation of Southern blot, 30 μg of genomic DNA samples from both transformed and nontransformed control plants were digested with *Eco*RI enzyme (Fermentas), which recognizes one restriction site within the T-DNA region. DNA samples were separated on 0.8% agarose gel, blotted on Hybond N + nylon membrane (Thermo Fisher Scientific) and hybridized with a nonradioactive DIG labeled probe (Sigma-Aldrich). For Northern hybridization, 15 μg of total RNA was used (Sambrook et al., 1989).

8.4.7 Analysis of Transgenic Progenies

Seeds that developed on PCR positive T_0 shoots following in-vitro flowering were harvested, dried under sunlight, and subjected for germination in plastic pots containing sterile peat moss along with their controls (nontransformed). The T_0 seeds germinated to give rise to T_1 plants; consequently, the T_1 plants produced T_2 plants. These raised progenies were characterized for the integration of *chitinase* gene by molecular analysis.

8.5. IN-VITRO REGENERATION OF SHOOTS

An efficient and reproducible in-vitro regeneration protocol is a prerequisite for the development of transgenic plants through genetic transformation (Gardner, 1993). Therefore, several attempts were made in establishing a suitable in-vitro regeneration method for microsperma types of lentils before conducting genetic transformation experiments. Among the various important crop plants, grain legumes have been considered recalcitrant due to their passiveness to in-vitro techniques (Mroginski and Kartha, 1984). Moreover, routine transformation protocols are not available in most of the grain legumes due to their poor in-vitro regenerating ability, especially via callus culture. Apart from this, lack of a compatible and efficient gene delivery system also hampers the development of desired transgenic plants in grain legumes. As a result, success of transformation in leguminous crops is limited (Nisbet and Webb, 1990). In spite of this fact, several reports are available on in-vitro regeneration of plantlets for a number of grain legumes, including pea (Schroeder et al., 1993), chickpea (Jayanand et al., 2003), common bean (Aragão et al., 2002), cowpea (Ikea et al., 2003), soybean (Hinchee et al., 1988), mung bean (Jaiwal et al., 2001), and peanut (Kartha et al., 1981). Although many legumes have been regenerated using tissue culture techniques, very few efficient regeneration protocols are presently available to exploit them in transformation experiments. Moreover, only a few reports are available on in-vitro regeneration of lentils, especially for microsperma types compared to other grain legume crops (Williams and McHugen, 1986; Saxena and King, 1987; Singh and Raghuvanshi, 1989; Malik and Saxena, 1992; Polanco and Ruiz, 1997; Gulati et al., 2001; Sarker et al., 2003b; Sevimoy et al., 2005; Sarker et al., 2012).

Three different types of explants, namely, cotyledonary node (CN), decapitated embryo (DE) and cotyledon-attached decapitated embryo (CADE), obtained from four microsperma varieties of lentils cultivated in Bangladesh namely, BM-1, BM-4, BM-5 and BM-6 were used to develop an in-vitro regeneration protocol compatible to *Agrobacterium*-mediated genetic transformation (Sarker et al., 2003b).

The explants of CN, DE, and CADE were cultured on MS medium supplemented with different concentrations and combinations of growth regulators such as BAP, Kn, GA_3, and NAA, as well as tyrosine as an amino acid supplement. Variable responses were observed toward the regeneration of shoots due to the application of different hormonal supplements with their wide range of concentrations and combinations. It was observed that CN explants produced multiple shoot buds on MS medium supplemented with 2.22 μM BAP, 2.32 μM Kn, 0.29 μM GA_3, and 30.35 μM tyrosine (Fig. 8.6) in all the varieties under investigation, only when there were pre-existing meristematic tissues in the explant. Maintaining these cultures resulted in initiation of multiple shoots from CN explants (Fig. 8.7).

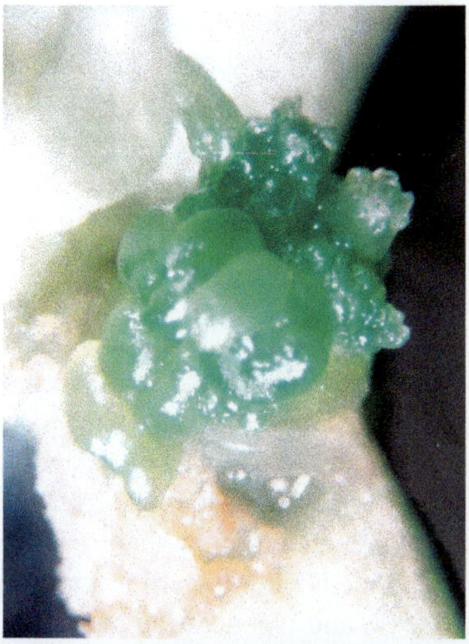

FIG. 8.6 Stereomicroscopic view of initiation of multiple shoot buds from cotyledonary node (CN) explant of BM-4 variety.

FIG. 8.7 Stereomicroscopic view of young multiple shoots as well as shoot bud developing from explant of cotyledonary node (CN) obtained from BM-4.

The highest number of healthy multiple shoots was obtained from CN explants (Table 8.1) on MS medium supplemented with 2.22 μM BAP, 2.32 μM Kn, 0.29 μM GA_3, and 30.35 μM tyrosine. In this combination, 2–6 shoots were found to develop from each explant, and their number was found to increase following the maintenance of these cultures in the same medium. It was observed that, at the base of the regenerated shoots, several new shoot buds of different sizes were also found to develop at later stages during the culture, which finally increased the number of regenerated shoots. These shoots were developed directly through organogenesis without the intervention of callus (Fig. 8.8). It was possible to develop elongated shoots from this type of explant using the aforementioned medium containing the same hormonal supplements. Fully developed healthy multiple shoots are presented in Fig. 8.9.

Next to cotyledonary node (CN), cotyledon-attached decapitated embryo explants (CADE) showed better responses for in-vitro shoot regeneration when the explants were cultured on MS medium containing 2.22 μM BAP, 2.32 μM Kn, 0.29 μM GA_3, and 30.35 μM tyrosine (Table 8.1). Decapitated embryo (DE) exhibited the lowest response toward multiple shoot regeneration. Use of embryo or modified embryo has been reported for in-vitro regeneration in peas (Schroeder et al., 1993) and chickpeas (Tewari-Singh et al., 2004; Jayanand et al., 2003) in achieving optimum in-vitro regeneration of shoots. It was reported that embryo attached to cotyledon responded better toward in-vitro regeneration of shoots in lentils (Halbach et al., 1998), and regeneration responses using embryo-derived explants were found to be almost identical in a number of similar experiments. It is worth mentioning here that the explants of cotyledonary node were the most responsive for the induction of multiple shoots through organogenesis in soybeans (Kaneda et al., 1997), peas (Jackson and Hobbs, 1990), chickpeas (Subhadra et al., 1998), and lentils (Warkentin and McHughen, 1993; Öktem et al., 1999; Mahmoudian et al., 2002) and mung bean (Saini et al., 2003).

A remarkable effect of tyrosine on the induction of shoots was recorded in the case of lentils. Tyrosine as an amino acid supplement has a positive role toward the development of in-vitro multiple of shoots in case of jute (Seraj et al., 1992) and wheat (Sarker and Biswas, 2002). Following these reports, several concentrations of tyrosine were used to enhance the regeneration capability of the explants. Among the various concentrations of tyrosine used, the numbers of regenerated shoots were found to be higher and shoots were healthy and elongated when 30.35 μM tyrosine was applied as a supplement to the regeneration medium. No remarkable variation was observed among the four different varieties of lentils growing in Bangladesh with regard to their multiple shoot regeneration capability.

8.6. IN-VITRO REGENERATION OF ROOTS

To obtain complete plantlets, induction of roots from the base of the regenerated shoot is an integral part of in-vitro regeneration studies. Development of

TABLE 8.1 Responses of Various Explants of Lentils Toward Multiple Shoot Formation When Cultured on MS Medium Supplemented With 2.22 μM BAP, 2.32 μM Kn, 0.29 μM GA_3, and 30.35 μM Tyrosine

Variety	Type of Explant	No. of Explant Inoculated	No. of Responsive Explants	% of Responsive Explants	Time to Start Regeneration (days)	No. of Shoot Per Explant (mean ± SD)
BM-1	CN	150	134	89.33	4–6	5.85 ± 1.25
	DE	113	49	44.59	12–16	1.45 ± 0.20
	CADE	110	80	72.72	5–7	4.25 ± 1.0
BM-4	CN	142	126	88.73	4–6	6.85 ± 1.75
	DE	124	58	46.0	12–16	3.45 ± 0.60
	CADE	130	97	74.6	5–7	5.75 ± 1.12
BM-5	CN	130	118	90.7	4–6	7.25 ± 2.27
	DE	160	68	42.5	12–16	2.95 ± 0.83
	CADE	265	197	74.3	5–7	4.95 ± 0.94
BM-6	CN	130	121	93.0	4–6	8.0 ± 2.05
	DE	105	42	40.0	12–16	3.1 ± 0.85
	CADE	280	216	77.1	5–7	6.15 ± 1.07

FIG. 8.8 Developing shoots from CN explants of BM-4 through organogenesis without the intervention of callus on MS medium containing 2.22 μM BAP, 2.32 μM Kn, 0.29 μM GA_3, and 30.35 μM tyrosine.

FIG. 8.9 Fully developed multiple shoots from CN explants on MS medium containing 2.22 μM BAP, 2.32 μM Kn, 0.29 μM GA_3 and 30.35 μM tyrosine in case of BM-1 variety.

roots has been tried from the in-vitro regenerated shoots that failed to develop roots spontaneously using a number of hormonal supplements. Formation of effective in-vitro roots has been reported to be difficult in microsperma varieties of lentils and is considered a major limiting factor in obtaining complete plantlets (Sarker et al., 2003b). However, there are a few earlier reports of rooting from regenerated shoots in lentils. Lentil shoots were found to produce roots on a sand bed in a mist chamber with an efficiency of 11% (Williams and McHugen, 1986). Moreover, the formation of roots from in-vitro regenerated shoots was achieved on a medium without hormonal supplements (Singh and Raghuvanshi, 1989). In a separate set of experiments higher percentage of rooting (95.35) in lentils was obtained by using nodal segments with an axillary bud when cultured in an inverted orientation (Fratini and Ruitz, 2003). However, rooting from in-vitro regenerated shoots of lentils was reported to be dependent on the nature and concentration of cytokinins (Polanco and Ruiz, 1997).

Previous reports further indicate that root induction in lentils can be achieved on a medium containing IAA and NAA (Polanco and Ruiz, 2001; Polowick et al., 2000). On the other hand, formation of roots was also reported on a medium containing IBA and NAA (Warkentin and McHughen, 1993). However, NAA was found to be more efficient than IBA for rooting in lentils (Ye et al., 2002), but 25% of in-vitro rooting in lentil was obtained by using 1.1 μM of IBA (Khawar et al., 2003). On the contrary, IAA, IBA, and NAA were used for rooting in other grain legumes including pea and mung bean (Tztzikas et al., 2004; Amutha et al., 2003).

Following the abovementioned reports, several hormonal combinations containing IBA and NAA were tried in order to induce roots in lentils. The explants with better shoot regeneration responses, namely, CN and CADE, were chosen for in-vitro root induction. Shoots regenerated from two different explants (CN and CADE) from all the varieties did not produce roots spontaneously. Therefore, induction of root growth was attempted by transferring freshly regenerated individual shoots of 2–4 cm lengths to MS media supplemented with 49.2–123.0 μM IBA. Root induction was observed after 10–15 days when shoots were cultured on MS medium supplemented with 98.4 μM or 123.0 μM IBA (Table 8.2). In these cases, the number of roots varied from 3 to 8 per shoot. However, these roots did not develop from the base of the shoots; rather, they developed slightly above the excision point (Fig. 8.10), and failed to integrate effectively with the shoot vascular system. Furthermore, freshly developed roots formed a callus when their tips were in contact with the medium (Fig. 8.11). This result was found to be identical to the studies (Polanco et al., 1988) where hormonal supplements of NAA or IAA were used.

Nonetheless, these reports demonstrate that rooting from in-vitro regenerated shoots in lentils remains a serious impediment for the development of a complete plantlet following use of an in-vitro regeneration system and moreover, successful in-vitro propagation of this plant species.

To overcome these problems of rooting, the technique of micrografting was applied (Gulati et al., 2001). However, lentil micrografting was found to be rather difficult, time-consuming, and complicated, and the rate of success varied. Thus, there was a need to develop some other efficient system, so that the recovery of regenerated complete plantlets could be possible.

TABLE 8.2 Effect of Different Concentrations of IBA in MS Medium on Root Induction From the Regenerated Shoots From BM-1 and BM-4 Varieties of Lentil

IBA Concentration in MS Media	Varieties of Lentil	No. of Shoots Inoculated	% of Rooting	Days to Root Induction	No. of Roots/Shoot
49.2 μM	BM_1	22	–	–	–
	BM_4	30	–	–	–
73.8 μM	BM_1	20	–	–	–
	BM_4	24	–	–	–
98.4 μM	BM_1	30	26.66	10–15	2–5
	BM_4	35	31.43	10–15	3–5
123.0 μM	BM_1	25	36.00	10–15	3–6
	BM_4	40	37.50	10–15	3–8

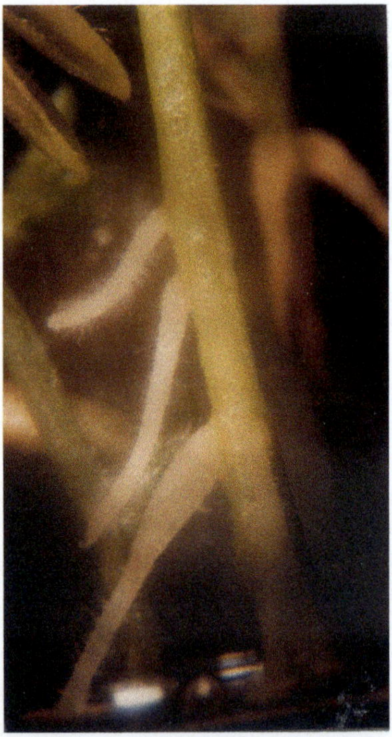

FIG. 8.10 Induction of adventitious roots from the body of the regenerated shoots slightly above the cut end of the regenerated shoots following treatment with a high concentration of IBA.

FIG. 8.11 Root induction from regenerated shoots on a medium containing a higher concentration of IBA; note when root tips were in contact with the medium, callus formation occurred (arrows), thus inhibiting further development of roots.

8.7. IN-VITRO FLOWERING AND POD FORMATION IN LENTILS

To overcome the limitations caused by poor root development from in-vitro regenerated shoots, an alternative regeneration system for lentils was developed, escaping the in-vitro root formation stage from regenerated shoots (Fig. 8.12). Under these circumstances, in-vitro flowering and fully developed seed formation can be considered a viable alternative to obtain plant progenies.

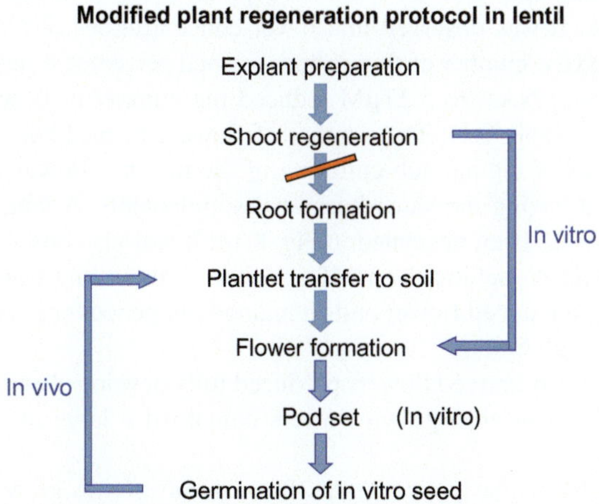

FIG. 8.12 Schematic diagram showing the modified plant regeneration protocol in lentils following the technique of in-vitro flowering and pod formation.

The necessary first step was to determine suitable media for in-vitro flowering and seed formation. It is understood that in-vitro flowering and pod production may significantly contribute toward the genetic improvement of lentils through biotechnological methods, particularly using *Agrobacterium*-mediated genetic transformation where the fertile seeds are expected to obtain from transformed shoots.

There is no report on in-vitro flowering and seed formation in lentils as well as in other grain legumes. However, few reports are available on in-vitro flowering in a number of plant species including citrus (Moss, 1969), cauliflower (Vandana et al., 1995), maize (Mandal et al., 2000), and bamboo (Nadgauda et al., 1990; Singh et al., 2000), among others, as well as in-vitro fruit development in tomato (Sheeja and Mandal, 2003). We are reporting here the successful development of in-vitro flowers and viable seed formation in microsperma lentil varieties (Sarker et al., 2012).

8.7.1 Development of In-Vitro Flowers

In-vitro-derived lentil shoots obtained from the explants of CN and CADE were transferred to MS or half strength of MS media supplemented with several concentrations and combinations of IAA or IBA plus NAA to induce in-vitro flowering and pod formation directly from in-vitro grown shoots. These shoots with developing flowers were maintained under in-vitro condition for self-pollination, fertilization, and seed setting. For these purposes, at least 500–600 in-vitro derived shoots were inoculated for each combination of hormonal supplements, and in-vitro flower buds were found to initiate within 7–10 days of culture (Fig. 8.13).

As presented in Table 8.3, the highest number (73.4%) of flowering shoots occurred on half-strength MS medium supplemented with 98.4 μM IBA and 2.69 μM NAA. It was observed that lower concentrations of NAA in the medium increased the number of flower buds formed per shoot, whereas increasing concentrations of NAA to 5.37 μM reduced the number of flower bud formation per shoot (Table 8.3). The number of flowers formed could be increased with two weeks of regular sub-culturing of the in-vitro, flower-bearing shoots to fresh media having the same hormonal supplements. A magnified view of in-vitro flower has been presented in Fig. 8.14. It was also observed that flower formation could be achieved on MS medium without any hormonal supplements, but these induced flower buds remained unopened for several weeks and failed to attain full bloom.

Several in-vitro-derived flowers produced fully developed anthers (Fig. 8.15) and anthers from in-vitro-grown flowers contained a large number of pollen grains (Fig. 8.16).

The viability of the pollen grains from these in-vitro-grown flowers was tested by FDA staining. Fluorescent microscopic observation indicated that approximately 80% of pollen grains were viable (Fig. 8.17). Self-pollinated pistils

FIG. 8.13 Development of several flowers from a regenerated shoot of BM-4 on half MS medium supplemented with 98.4 μM IBA and 2.69 μM NAA.

were also examined to monitor in-situ growth of the pollen tubes within the stylar tissue. Callose was found to deposit within healthy pollen tubes, and such pollen tubes were observed using fluorescent microscope with the help of aniline blue stained pistil obtained from the in-vitro flowers (Fig. 8.18). After such self-pollination, the ovules were found to develop into seeds (Fig. 8.19).

Despite the lack of reports on in-vitro flowering and pod formation in lentil as well as in other grain legumes, a limited number of studies have been conducted in ornamental plants, as in the case of in-vitro flowering in bitter melon (Wang et al., 2001), *Panax ginseng* (Tang, 2000), *Murraya paniculata* (Jumin and Ahmed, 1999), *Basilicum polystachyon* (Amutha et al., 2008), and *Rosa hybrida* (Kanchanapoom et al., 2009). Flowering is considered a complex process regulated by a combination of environmental and genetic factors (Bernier, 1988; Weigel, 1995; Weller et al., 1997). On the other hand, phytohormones can also influence diverse developmental processes pertaining to in-vitro flower formation (McCourt, 1999). Previous reports indicated that in-vitro flowering in ornamental and medicinal plants had been achieved on MS medium containing a variety of hormonal supplements including BAP and Fe^{2+} (Wang et al., 2001; Jumin and Ahmed, 1999), GA_3 and NAA (Tang, 2000), BAP and Kn (Amutha et al., 2008), and BAP and GA_3 (Kanchanapoom et al., 2009).

TABLE 8.3 Effects of Different Concentrations of Auxins on In-Vitro Flowering and Pod Setting

Media Composition	No. of Shoots Inoculated	% of Shoots With Flowers	% of Flowering Shoots With Pod Set	Range of Flowers Per Shoot	Range of Pods Per Shoot
MS without auxins	570	10.52	0	0	0
MS + 49.2 μM IBA	580	6.89	0	1	0
MS + 49.2 μM IBA + 5.37 μM NAA	500	60.00	25.00	1–5	1
MS + 57.08 μM IAA + 5.37 μM NAA	510	17.64	0	1–2	0
MS + 49.2 μM IBA + 2.69 μM NAA	640	68.75	22.73	1–8	1
MS + 57.08 μM IAA + 2.69 μM NAA	450	33.33	20.00	1–2	1
MS + 98.4 μM IBA	405	14.81	25.00	1–3	1
MS + 98.4 μM IBA + 5.37 μM NAA	510	41.17	28.57	1–3	1
MS + 114.1 μM IAA + 5.37 μM NAA	312	9.62	0	1–2	0
MS + 98.4 μM IBA + 2.69 μM NAA	540	63.38	13.04	1–10	1–2
MS + 114.16 μM IAA + 2.69 μM NAA	525	42.85	6.66	1–4	1
½ MS + 98.4 μM IBA + 2.69 μM NAA	585	73.35	57.57	1–6	1–4
½MS + 114.16 μM IAA + 2.69 μM NAA	525	48.57	41.17	1–4	1–3

FIG. 8.14 Enlarged view of developing in-vitro flower in the case of the BM-4 variety of lentil (×10).

FIG. 8.15 Magnified view of fully developed in-vitro flower of BM-6 showing different floral parts including pistils and anthers (×12).

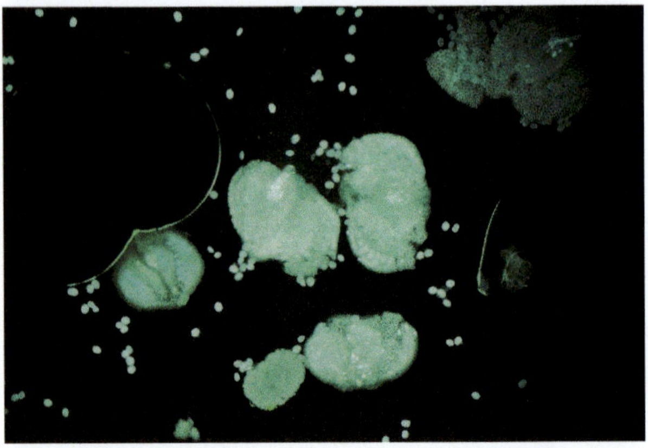

FIG. 8.16 Anthers from in-vitro-grown flowers with large number of pollen grains observed under fluorescence (×12).

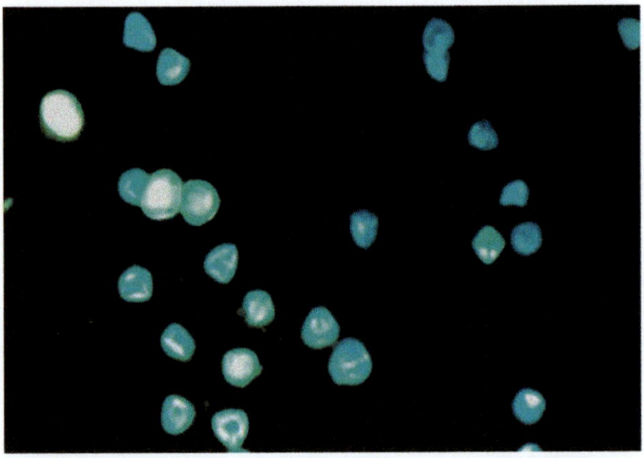

FIG. 8.17 Fluorescent microscopic view of pollen grains following FDA staining showing several viable pollen grains (×20).

8.7.2 In-Vitro Pod and Seed Development

The flower bearing shoots were maintained by fortnightly subcultures to the same medium until in-vitro pods had fully matured. About 12–20 days after flowering, pods were found to develop from healthy flowers (Fig. 8.20). The best response regarding the number of pods per shoot was obtained on half-strength MS medium supplemented with 98.4 μM IBA and 2.69 μM NAA. About 58% of the flowers produced pods, and the range of pods per shoot was 1–4 (Table 8.3). The morphology of the in-vitro-raised flowers was indistinguishable from flowers grown in control fields, including the number of seeds set per pod.

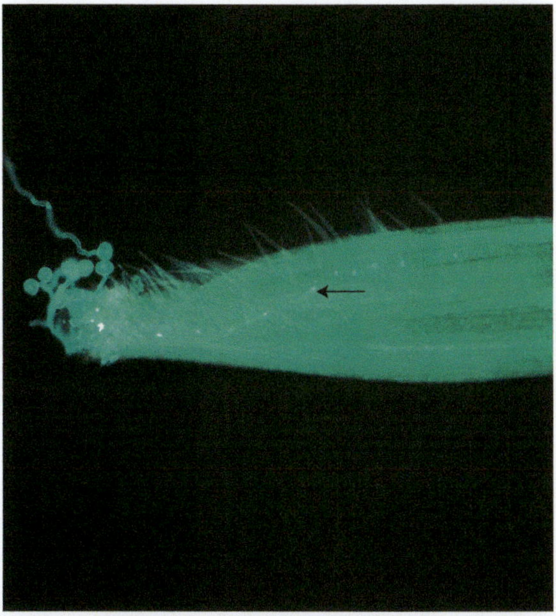

FIG. 8.18 Fluorescence micrograph of callose invested healthy pollen tubes (arrows) within the tissue of the pistil from the in-vitro-developed flowers (×20).

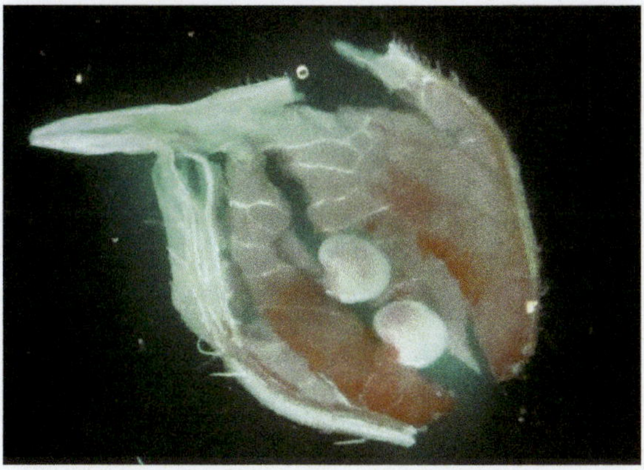

FIG. 8.19 Fluorescent micrograph showing the developing seeds within the ovary following self-pollination (×10).

Because half-strength MS medium successfully induced flowering and pod set, the comparative capability of the two different explants for flower development and pod formation was examined using this media only. In all four varieties of lentil, shoots derived from cotyledonary nodes showed a better response toward in-vitro flower formation, without any remarkable variation between

FIG. 8.20 Formation of in-vitro pods from regenerated shoots on ½ MS medium containing 98.4 μM IBA plus 2.69 μM NAA.

two different types of explants for in-vitro pod formation (Table 8.4). It was also noted that there was no significant difference in number of flowers formed per shoot among the four varieties of lentils.

The procedure of in-vitro flower and pod formation described here can be utilized for the successful recovery of in-vitro-regenerated lentil plantlets, as well as being a critical step toward the development of a transformation protocol required for a lentil improvement program.

8.7.3 In-Vitro Germination of Seeds and Establishment of Plantlets Into Soil

The in-vitro-grown pods were allowed to grow fully within 20 to 30 days under in-vitro conditions (Fig. 8.21). Pods were then harvested and dried under sunlight (Fig. 8.22). Seeds were then surface sterilized and germinated on water agar medium. These germinated seedlings were then transplanted into soil for further development, and it was observed that most of these transplanted plants eventually developed flowers and seeds (Fig. 8.23). At the same time, seeds harvested from in-vitro-raised pods were also directly sown in plastic pots containing sterilized peat moss. Pots were then kept in a culture room for two

TABLE 8.4 Comparison of the Effects of Different Sources of Explants for In-Vitro Flowering and Pod Formation Cultured on Half-Strength MS Medium Supplemented With 98.4 μM IBA + 2.69 μM NAA

Explant Type	No. of Shoots Inoculated	No. of Shoots With Flowers	% of Shoots With Flowers	% of Shoots With Pod	Days to Initiate Flowers	Days to Pod Set	Range of Flowers Per Shoot	Range of Pods Per Shoot
Cotyledonary node	420	340	77.00	50.00	12–15	15–20	1–8	1–4
Cotyledon-attached decapitated embryo	330	210	64.66	47.62	8–10	12–15	1–6	1–5

FIG. 8.21 Completely mature pods under in-vitro condition.

FIG. 8.22 Harvested, in-vitro-derived mature pods.

FIG. 8.23 Plantlet developed from in-vitro-grown seeds transferred to soil producing flowers and pods.

weeks and then further maintained in a net house for flower and pod development. In this experiment, it was observed that seeds that directly sown in plastic pots containing sterilized peat moss showed better response toward germination compared to the seeds that placed on water agar medium.

8.8. *AGROBACTERIUM*-MEDIATED GENETIC TRANSFORMATION USING MARKER GENES

Conventional breeding techniques appeared to be inadequate in producing desired results for the improvement of the agronomic performance of lentils, particularly toward the development of stress tolerance. It is believed that conventional breeding methods need to be complemented by biotechnologically based techniques for the improvement of presently available lentil varieties. Therefore, plant genetic engineering can be considered one of the methods of choice for the introduction of new traits in lentil for its improvement.

Establishment of an effective genetic transformation protocol is required for the integration of desired genes in lentil varieties. A number of methods are available for genetic transformation of plants, including *Agrobacterium*-mediated genetic transformation, microprojectile bombardment, electroporation, sonication, chemical transformation, etc. Generally, *Agrobacterium*-mediated genetic transformation has been considered the most convenient and cost-effective technique. It is the most popular method for genetic transformation because of high co-expression of introduced genes, potentially producing low copy numbers and preferential integration into active transcribable regions (Birch, 1997; Gheysen et al., 1998). Therefore, *Agrobacterium*-mediated genetic transformation technique has been followed to insert desired genes into microsperma varieties of lentils.

Lentils are considered an underexploited crop because limited efforts have been made so far to improve lentils through plant genetic engineering in comparison to other pulse crops. It is worth mentioning here that *Agrobacterium*-mediated genetic transformation has been successfully used in the case of a number of grain legumes including soybeans (Hinchee et al., 1988; Meurer et al., 1998), chickpeas (Fontana et al., 1993; Kar et al., 1996; Krishnamurthy et al., 2000), peanuts (McKently et al., 1995; Krishna et al., 2015), pigeon peas (Surekha et al., 2005), etc.

Nevertheless, few reports are available, and they mostly concern the development of a genetic transformation protocol in case of macrosperma-type lentil plants using *Agrobacterium*-mediated genetic transformation. A variety of lentil explants, including shoot apices, epicotyls, cotyledonary nodes, nodal segments, embryonic axes, and roots were exploited for *Agrobacterium*-mediated genetic transformation (Warkentin and McHughen, 1991, 1992, 1993; Mahmoudian et al., 2002; Öktem et al., 1999; Lurquin et al., 1998; Akcay et al., 2009). Apart from this, particle bombardment method of transformation was explored in lentil (Gulati et al., 2002), and the in planta method of transformation was also exploited (Chowrira et al., 1996). None of these studies produced effective, stable, transformed lentil plants using marker genes, and in these investigations, no specific gene of interest was targeted.

However, studies on the *Agrobacterium*-mediated genetic transformation using the microsperma varieties of lentils were initiated (Sarker et al., 2003a). Following that, study efforts were made for the development of a complete and efficient transformation protocol for microsperma lentil varieties through *Agrobacterium*-mediated genetic transformation. For this purpose, *Agrobacterium tumefaciens* strain LBA4404 harboring binary plasmid pBI121, containing *GUS* and *npt*II genes were used. This gene construct was used particularly for the optimization of transformation protocol and to select the suitable lentil explant type for transformation. *GUS* (β-glucoronidase) gene was used as a screenable marker and *npt*II gene as a selectable marker conferring kanamycin resistance. Generally, transient assays for such marker gene (*GUS*) were routinely performed as a preliminary step to identify the conditions required for successful transfer of desired genes.

Three microsperma varieties of lentil, namely, BARI Masur-4 (BM-4), BARI Masur-5 (BM-5), and BARI Masur-6 (BM-6) were used for *Agrobacterium*-mediated transformation. For these purposes, three different types of embryo-derived explants, namely, cotyledonary node (CN), decapitated embryo (DE), and cotyledon-attached decapitated embryo (CADE), were used.

The *Agrobacterium*-mediated genetic transformation procedure is believed to be influenced by several factors (Mansur et al., 1993). Factors that influence successful transformation, such as optical density (OD) of *Agrobacterium* suspensions, incubation, and co-cultivation periods, were optimized. During optimization of various regulatory factors, a bacterial suspension having an OD of 1.0 with 45 min of incubation and 3 days of cocultivation was found to be optimum for transformation of lentil explants.

It was reported that transgenic chickpea plants were obtained by incubating mature embryo explants for 20 min followed by 3 days of cocultivation (Krishnamurthy et al., 2000). Transformed plantlets of chickpea were developed by employing the same cocultivation period of 3 days, but the incubation period for the explants in bacterial suspension was maintained for 1–2 h (Tewari-Singh et al., 2004).

8.8.1 Responses of Various Explants of Lentils Toward *Agrobacterium*-Mediated Genetic Transformation

The transformation ability of various explants, namely, DE, CN, and CADE of BM-4, BM-5, and BM-6 lentil varieties was monitored through the expression of the *GUS* gene, following incubation for a period of 45 min with *Agrobacterium* strain LBA4404 having an OD of 1.0 and cocultivation for 72 h. *GUS* expression was detected by histochemical assay, and such expression was characterized by the formation of an indigo blue color within the transformed cells of the infected explants. Control explants were always maintained in each set of experiment and were subjected to a *GUS* histochemical assay in order to understand the difference between transformed and nontransformed tissue. A good number of cocultured explants showed positive to *GUS* staining. *GUS*-positive regions were visualized mostly at the peripheral area of the cut surfaces as well as within the internal tissues of various explants (Figs. 8.24–8.26).

DE explants showed the best response toward transformation with LBA4404 strain, and the percentage of GUS-positive DE explants was 86.95 in the case of the BM-6 variety of lentils. Next to DE explants, CADE explants showed better responses toward transformation, and in this case, the percentage of GUS-positive explants was found to be 78.18. CN explants exhibited the lowest responses toward *GUS* expression among all the explants studied for all the varieties (Table 8.5).

Although transformation efficiency was comparatively low in CN explants, its regeneration capacity was found to be better among the three different types of explants studied. Therefore, this explant was included for further transformation experiments. On the other hand, in spite of better transformation efficiency

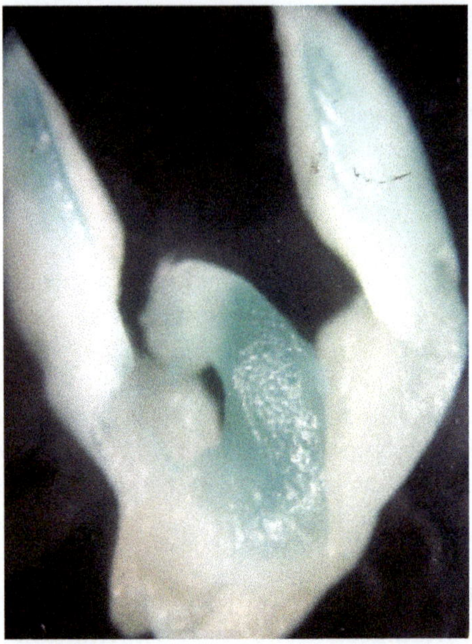

FIG. 8.24 Histochemical localization of GUS activity in the explants of CN from BM-4 (×30).

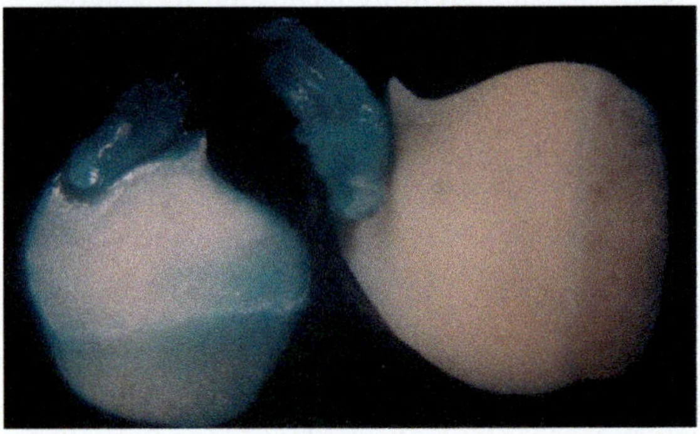

FIG. 8.25 Histochemical localization of GUS activity in case of CADE explants of BM-5 (×20).

exhibited by DE, further transformation experiments were not carried out using this type of explant, because the regeneration capability of this explant was found to be very low. Thus, most of the transformation experiments were carried out with only CN and CADE explants.

8.8.2 Regeneration and Selection of Transformed Shoots

To obtain regeneration of shoots, cocultivated explants were washed with 300 mg/L ticarcillin for 15 min and then transferred to MS medium containing

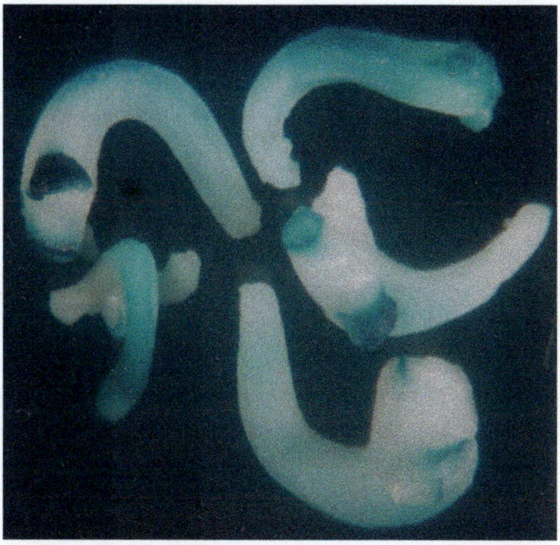

FIG. 8.26 Histochemical localization of GUS activity in the explants of DE from BM-4 (×20).

TABLE 8.5 Responses of Various Explants of Lentil Toward GUS Histochemical Assay Following Cocultivation With *Agrobacterium* Strain LBA4404

Variety	Explants	No. of Explants Assayed for GUS	No. of GUS +ve Explants	% of GUS +ve Explants
BM-4	CN	95	53	55.79
	CADE	90	70	77.78
	DE	85	72	84.70
BM-5	CN	146	47	32.19
	CADE	120	91	75.83
	DE	126	105	83.33
BM-6	CN	150	58	38.67
	CADE	110	86	78.18
	DE	138	120	86.95

CN, cotyledonary node; *CADE*, cotyledon-attached decapitated embryo; *DE*, decapitated embryo.

2.22 μM BAP, 2.32 μM Kn, 0.29 μM GA3, and 30.35 μM tyrosine for shoot proliferation via organogenesis (Das et al., 2012). In addition to these supplements, the shoot-regeneration media was also supplemented with 100 mg/L ticarcillin for the elimination of all unwanted bacteria, but no selectable agents were applied immediately after cocultivation.

The shoots developed from both infected and noninfected (control) CN and CADE explants were transferred to selection medium. To determine the optimum selection level, noninfected explants (those which served as negative control) were subjected to selection pressure. For this experiment, after the initiation of regeneration, the explants were transferred to a suitable regeneration medium containing 50 mg/L kanamycin. Then, the concentration of kanamycin was increased gradually in the following manner: 50, 100, 150, and 200 mg/L.

From this experiment, it was revealed that, with the increase of kanamycin concentration, the percentage of survived shoots was found to decrease. Due to the effect of kanamycin, the shoots first became albino (Fig. 8.27), and finally failed to survive.

The effects of different concentrations of kanamycin on regenerated shoots from control explants were examined. In this study, it was observed that all the noninfected explants (negative control) failed to survive in the presence of 200 mg/L kanamycin within 15 days of inoculation (Fig. 8.28). Therefore, shoots developed from infected explants surviving in regeneration medium with 200 mg/L kanamycin for more than 15 days were considered to be transformed.

The regenerating shoots with 2–3 leaves were subcultured on the same media, but with kanamycin for selection of transformed shoots. Initiation of multiple shoots was found to be retarded in the presence of kanamycin. Therefore, a preculture period and a delayed selection with kanamycin were followed in obtaining regeneration from explants with higher transformation efficiency. Similar observation during selection of transformed shoots was also reported in other plant species such as alfalfa, chickpeas, peanuts, and lentils (McHughen et al., 1989; Pezzoti et al., 1991; Kar et al., 1996; Sarker et al., 2003a).

FIG. 8.27 Albino shoots developed due to the effect of 100 mg/L of kanamycin during optimization of selection of transformed shoots using different concentrations of kanamycin.

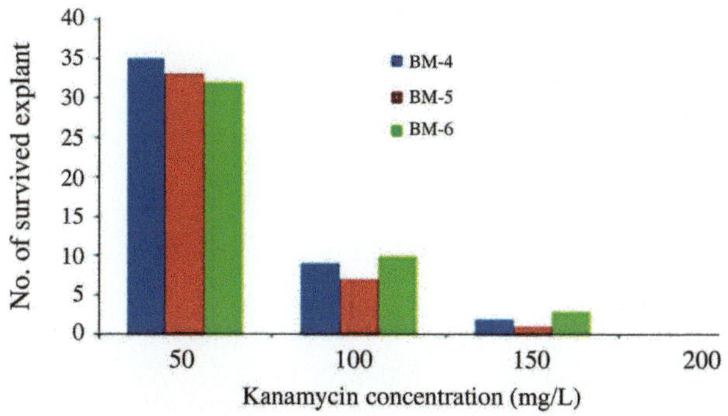

FIG. 8.28 Effect of various concentrations of kanamycin on developing shoots regenerated from noninfected CADE explants.

For successful transformation, a proper and efficient selection procedure is required in obtaining desired transformed shoots/plantlets. Selection of the transformed shoots was carried out by gradually increasing the concentration of kanamycin upto 200 mg/L. To select the transformed shoots, initially, 50 mg/L kanamycin was used as the selection pressure. After 14 days, only green and healthy shoots were subcultured on fresh regeneration medium with 100 mg/L kanamycin. Then, the kanamycin concentration was raised to 150 mg/L in the third subculture, and then 200 mg/L in the fourth subculture. During each subculture, the albino and deep brown dead shoots were discarded, and only green shoots were subcultured on fresh medium containing the next higher concentration of kanamycin. From this experiment, it was revealed that with the increase of kanamycin concentration the percentage of survived shoots were found to decrease (Table 8.6). Finally, those shoots that survived on regeneration medium with 200 mg/L kanamycin for more than 15 days and remained green and healthy were selected as possible transformed shoots.

It was observed that none of the CN explants derived shoots were able to continue their growth in presence of higher concentration (200 mg/L) of kanamycin, whereas a few CADE-explants-derived shoots were recovered after such selection (Fig. 8.29).

In a series of experiments, a total of 31 kanamycin resistant shoots out of approximately 1896 infected explants in case of BM-4, 12 out of 570 in case of BM-5 and 22 out of 1020 in case of BM-6 were recovered in final selection medium with LBA4404 strain of *Agrobacterium* containing binary plasmid pBI121. Therefore, the frequency of recovery of putatively transformed shoots was about 1.63 in the case of BM-4, 2.10 in the case of BM-5, and 2.15 in the case of BM-6 (Table 8.6).

Identical concentrations of kanamycin and a procedure for the recovery of transgenic shoots in the cases of BM-2 and BM-4 lentil varieties using decapitated embryo explants, were utilized (Sarker et al., 2003a). In chickpea, two concentrations of kanamycin (25 and 50 mg/L) were applied for the selection

TABLE 8.6 Effect of Kanamycin on Developing Shoots During Selection of Transformed Shoots

Variety	Explant	No. of Explants Infected	No. of Shoots Inoculated	No. of Survived Shoots in Regeneration Media With Kanamycin (mg/L)				% of Survive Shoots
				50	100	150	200	
BM-4	CN	480	1066	1010	323	53	0	0
	CADE	1896	3125	3072	1011	307	31	1.63
BM-5	CN	410	1020	940	234	13	0	0
	CADE	570	1162	1125	460	256	12	2.10
BM-6	CN	400	1009	920	201	22	0	0
	CADE	1020	2982	2062	490	56	22	2.15

CN, cotyledonary node; *CADE*, cotyledon-attached decapitated embryo.

FIG. 8.29 Putatively transformed shoots (green in color) of BM-4 survived on the selection medium containing 200 mg/kanamycin, while nontransformed shoots are albino in nature.

of transformed shoots (Kar et al., 1996), while increased concentrations of kanamycin were utilized gradually from 50 to 100 mg/L in obtaining transformed peanut plantlets (Eapen and Goerge, 1994).

Histochemical *GUS* assay (Jefferson et al., 1987) was used to detect the expression of the *GUS* gene in developing shoots and leaves at the time of each subculture in selection medium. Such assays demonstrated the presence of blue-colored zones on selected shoots and leaves (Fig. 8.30).

Chimeric expression of the *GUS* gene (Fig. 8.31) was also observed in a number of transformed shoots. Moreover, all shoots that survived on the final

FIG. 8.30 Putatively transformed shoots of lentil showing the expression of *GUS* gene.

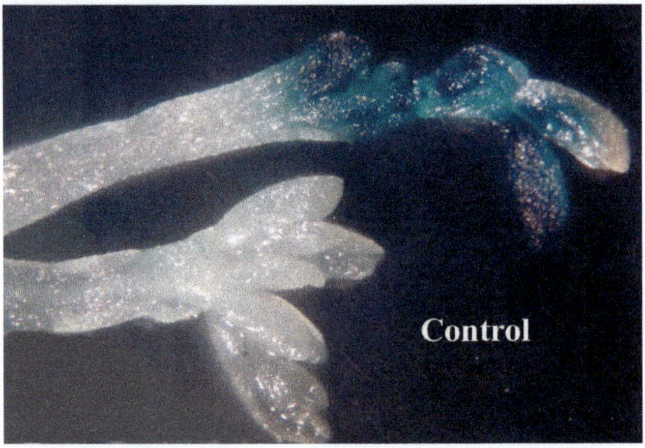

FIG. 8.31 Histochemical localization of *GUS* activity in developing shoots and leaves showing chimeric-type expression. The control shoot (bottom) without the expression of *GUS* gene.

selection pressure did not show positive responses toward GUS expression. A number of researchers reported this kind of phenomenon in other crops, including *Carthamus* (Ying et al., 1992); expression of the *GUS* gene was found only in some kanamycin-resistant potato calli (Ottaviani et al., 1993). These observations indicated that the expression of the GUS activity in the regenerated shoots was not directly correlated with kanamycin resistance. The lack of GUS expression in kanamycin resistant shoots may be due to alternation or loss of *GUS* gene resulting from rearrangement of the coding sequence or methylation of the gene (Battraw and Hall, 1990; Ottaviani et al., 1993).

From our previous experiments, it was observed that the in-vitro regenerated shoots of microsperma lentil varieties failed to develop effective in-vitro roots, even on the medium containing higher concentration of auxins (Sarker et al., 2003a). Because in-vitro rooting has appeared to be a limitation in obtaining complete plantlets in lentils, in-vitro flowering and seed formation were induced in transformed lentil shoots (GUS positive) with a view to recovery of the transgenic lentil progenies.

8.8.3 In-Vitro Flowering and Pod Formation Using Transformed Shoots

For the recovery of transgenic lentil plantlets, the technique of in-vitro flowering was applied as an alternative to in-vitro rooting. In-vitro flowering was applied in order to obtain fertile seeds from transformed shoots. For this purpose, the shoots that survived following selection were allowed to develop in-vitro flowers in the flower developing culture medium. A series of experiments demonstrated that half-strength MS medium supplemented with 98.4 μM IBA and 2.69 μM NAA showed the best response toward in-vitro flowering and pod formation. However, it was observed that in higher concentrations of kanamycin (100 mg/L), about 50% of flower buds failed to open, and in 200 mg/L of kanamycin, no flower bud formation was observed. Therefore, shoots that survived in the selection medium were separated and transferred to half-strength of MS medium containing 98.4 μM IBA and 2.69 μM NAA and 50 mg/L kanamycin. After 2–3 weeks, in-vitro flower formation was observed in the healthy shoots recovered through kanamycin selection (Fig. 8.32). It was found that, out of 43 transformed shoots, only 11 responded to flowering, and the maximum number of flowers per shoot was 3. It was also found that after 12–15 days, 5 out of these 11 flowering shoots produced viable and healthy pods (Fig. 8.33) under in-vitro conditions.

Healthy in-vitro seeds developed from the selected shoots were harvested and dried under sunlight. Seedlings developed from these in-vitro raised seeds were successfully transplanted into soil for development of transformed progenies (Fig. 8.34).

Responses of the CADE-derived shoots from three lentil varieties were examined to study their transformation efficiency. Such shoots were exposed to kanamycin for the selection of transformants, and then, the putatively transformed shoots were allowed to develop in-vitro flowers and pods, and their transformation efficiency was recorded following the required polymerase chain reaction (PCR) analysis using a marker gene (Table 8.7).

FIG. 8.32 Formation of in-vitro flower on the shoot recovered through kanamycin selection.

FIG. 8.33 Fully developed in-vitro pod from putatively transformed shoot.

FIG. 8.34 Lentil seedling developed from in-vitro-raised, transformed seed was successfully transplanted into soil for further development of progenies.

TABLE 8.7 Responses of CADE-Derived Selected Shoots Toward In-Vitro Flowering and Pod Formation and Their Transformation Efficiency Using a Marker Gene

Variety	No. of Shoots Subjected to Selection	No. of Shoots Survived in 200 mg/L Kanamycin	No. of Shoots With Flower	No. of Shoots With Pod	No. of PCR +ve Plants	Transformation Efficiency (%)
BM-4	3072	31	6	2	0	0
BM-5	1125	12	4	2	2	0.18
BM-6	2062	22	10	3	4	0.19

8.8.4 Analysis of Transgene Integration

The transgenic nature of the putative transformed shoots was confirmed by amplification of *GUS* and *npt*II genes present within the genomic DNA of the transformed shoots. For this purpose, total genomic DNA was isolated from the leaves of putative transformants and control lentil shoots. This isolated DNA was subjected to PCR for the amplification of *GUS* and *npt*II genes. At the same time, plasmid DNA of pBI121 as maintained in the genetically engineered *Agrobacterium*, having *GUS* and *npt*II genes (marker gene construct), was isolated and used as positive control during PCR analysis. Gene-specific primers were used for the amplification of *GUS* and *npt*II genes. The amplified DNA was visualized through agarose gel electrophoresis (Figs. 8.35 and 8.36). These findings indicated the presence of transgene within the genomic DNA of plantlets/shoots.

8.9. GENETIC TRANSFORMATION USING ANTIFUNGAL GENE CONSTRUCTS

Agrobacterium tumefaciens-mediated genetic transformation was carried out to integrate fungal-disease-resistant genes into microsperma lentil varieties using two gene constructs, namely, EHA105/pGII-pSOUPvstN-chitin (Construct II) and LBA4404/pCAMBIA2300 enh35SAFP (Construct III). This study was performed following the transformation protocol developed by exploiting the marker genes. In this case, Construct II contained the *bar* gene as a selectable marker that conferred phosphinothricin (PPT) resistance and chitinase as an antifungal gene (the gene of interest), whereas Construct III contained *npt*II gene as a selectable marker that conferred kanamycin resistance and an antifungal protein gene (*AFP*) as the second gene of interest.

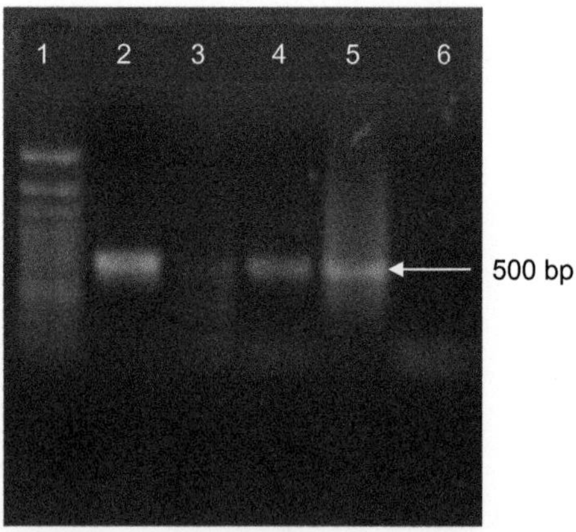

FIG. 8.35 PCR amplification of *GUS* gene from transformed lentil shoots. Lane 1: 1 kb ladder; Lane 2: plasmid DNA of pBI121 as positive control; Lane 3: negative control; Lanes 4 and 5: genomic DNA of transformed shoots; Lane 6: water control.

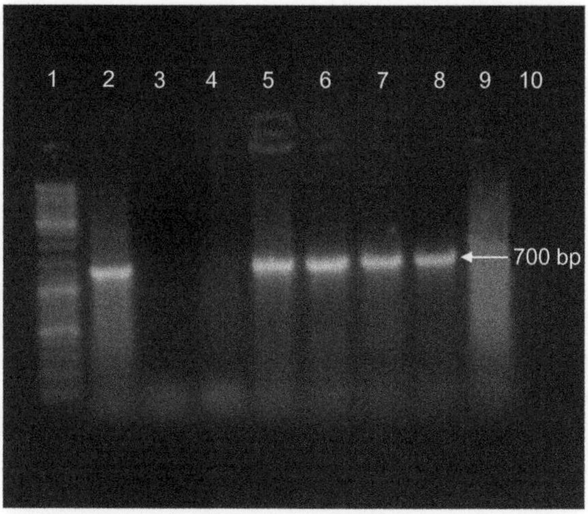

FIG. 8.36 PCR amplification of *npt*II gene from transformed shoots of lentil. Lane 1: 1 kb ladder; Lane 2: plasmid DNA of pBI121 as positive control; Lanes 3–8: genomic DNA of putative transformed shoots; Lane 9: negative control; Lane 10: water control.

8.9.1 Transformation With EHA105/pGII-pSOUPvst N-chitin

Transformation experiments were carried out only with CADE explants from BM-4, BM-5, and BM-6 lentil varieties. Different parameters influencing transformation efficiency using *Agrobacterium tumefaciens* strain EHA105 were optimized. Based on the results of a series of experiments, it was revealed that 3 days of cocultivation period with optical density of 1.0 is optimal for transformation with this gene construct.

Following cocultivation for 3 days, explants were washed with 300 mg/L ticarcillin for 10 min and then transferred to MS medium supplemented with 2.22 μM BAP, 2.32 μM Kn, 0.29 μM GA_3, and 30.35 μM tyrosine, as well as with 100 mg/L ticarcillin for the regeneration of shoots, as well as elimination of all unwanted bacterial outgrowth.

Then, for the selection of transformed shoots phosphinothricin (PPT) was used as selective agent as this gene construct contained *bar* gene which is resistant to PPT. Selection of transformed shoots was achieved by gradually increasing the concentrations of PPT, starting from 1.0 to 2.0 mg/L. The regenerating shoots with young leaves were subcultured on the same media (2.22 μM BAP, 2.32 μM Kn, 0.29 μM GA_3, and 30.35 μM tyrosine) containing 1.0 mg/L PPT for selection. After 14 days, only green and healthy shoots were subcultured on fresh medium with 1.5 mg/L PPT. To select transformed shoots, the PPT concentration was raised to 2.0 mg/L in the third subculture, and then again in 2.0 mg/L in the fourth subculture. During each subculture, deep brown dead shoots were discarded, and only green shoots were subcultured on fresh medium containing the next higher concentration of PPT. From this experiment, it was noticed that with increased PPT concentration, the percentage of survived shoots decreases. Among the several different concentrations tried, 2.0mg/L PPT was found to be effective in selecting the transformed shoots.

Finally, shoots that survived on regeneration medium containing 2.0mg/L PPT for two rounds (each for 15days) and remained green and healthy were selected as possible transformed shoots (Fig. 8.37).

The putative transformed shoots were exposed to in-vitro flowering and pod formation medium for the development of transformed progenies. Shoots that survived in the PPT selection medium were transferred to half-strength MS medium supplemented with 98.4μM IBA and 2.69μM NAA for in-vitro flowering and subsequent development of pods.

After 2–3weeks, in-vitro flower formation was observed on the healthy shoots (Fig. 8.38), and eventually, pod formation occurred from the shoots that were growing on flower-inducing medium. The overall results for the recovery of transformed shoots through PPT selection and ability of these shoots toward in-vitro flower and seed formation are presented in Table 8.8.

Molecular analysis was carried out to confirm the presence of *bar* and *chitinase* genes in putative transformants. Pods that developed on PCR-positive T_0 shoots were harvested. The T_0 seeds were germinated and gave rise to T_1 plants and subsequently T_2 plants were developed from T_1 seeds. Genomic DNA was isolated from T_1 and T_2 plants for required PCR amplification. Using proper primers, PCR amplification for the *chitinase* gene was done for DNA samples obtained from the respective T_1 and T_2 plants. The results of PCR amplification of T_1 and T_2 plants are shown in Figs. 8.39 and 8.40.

FIG. 8.37 Putative transformed green and healthy shoots recovered through selection using 2.0mg/L PPT.

FIG. 8.38 Development of in-vitro flower on the healthy shoots obtained through PPT selection.

8.9.2 Transformation With LBA4404/pCAMBIA2300 enh35S AFP

Further experiments on plant genetic transformation were conducted to integrate another antifungal protein (AFP) gene into the genome of lentil varieties. For this purpose, transformation experiments were carried out following the protocol developed for microsperma lentils using the marker genes as described in Section 8.8. In this set of experiments, CADE explants from the BM-6 variety of lentil and the *Agrobacterium tumifaciens* strain LBA4404 containing the plasmid vector pCAMBIA2300 with *AFP* gene were utilized. Earlier various parameters influencing transformation efficiency were optimized.

In this case, the *npt*II gene was used as a selectable marker conferring kanamycin resistance. Therefore, selection of transformed shoots was carried out using optimum concentration of kanamycin. The shoots which survived in 200 mg/L kanamycin containing regeneration medium were regarded as putative transformants. All the five putative transformed shoots were subjected to PCR analysis with the specific primers to detect the presence of *npt*II and *AFP* genes. From this experiment, it was observed that two out of five putative transformed shoots developed flower and pods.

Reverse Transcription PCR (RT-PCR) was performed on cDNAs synthesized from the total RNAs of these five PCR positive plants. The same base sequences are used as primers that were used in regular PCR reactions. The *npt*II and *AFP* primers amplified the expected 750 bp and 190 bp products, respectively, from the total cDNAs of four PCR-positive shoots (Figs. 8.41 and 8.42).

TABLE 8.8 Effect of PPT on Developing Shoots Toward the Selection of Transformants

Variety	Total No. of Explant Infected	No. of Explant Transferred to Regeneration Media	No. of Shoot Survived (mg/L PPT)			No of Shoots Transferred to Flowering Media	% of Shoots Survived Following Selection
			1.0	**1.5**	**2.0**		
BM-4	1390	1303	107	15	5	5	0.36
BM-5	2040	1980	123	21	7	7	0.34
BM-6	4680	4540	243	33	17	17	0.36

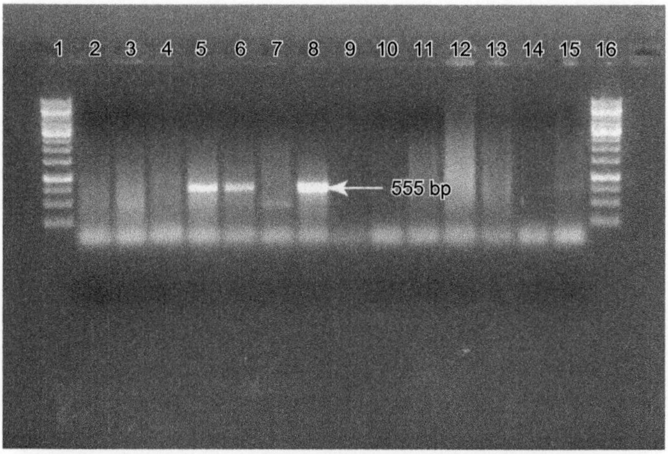

FIG. 8.39 PCR amplification of the *chitinase* gene from T_1 plants (Lanes 5 and 6), Lane 8 is a positive control, Lanes 2, 3, 4, 7, 9, 12, 13, 14, and 15 are nontransformed plants; Lane 10: water control; Lane 11: negative control; Lanes 1 and 16 are 1 kb ladders.

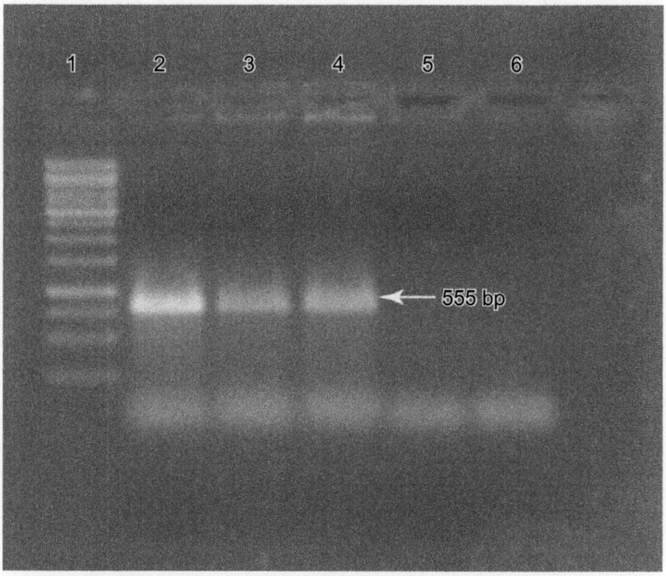

FIG. 8.40 PCR amplification of *chitinase* gene from T_2 plants, note that Lanes 3 and 4 were obtained from T_1 plant (Lane 5 of Fig. 8.3); Lane 2: positive control; no signal from negative control (Lane 6); water control (Lane 5); Lane 1: 1 kb ladder.

Southern blot analysis was also carried out using five putative transformed shoots. Among these five shoots, three did not show any signal, while one of them had a single insertion, and another had double insertions (data not shown). Results of Southern hybridization confirmed the integration of the transgenes into the genome of transgenic lentil lines. Further molecular analysis was

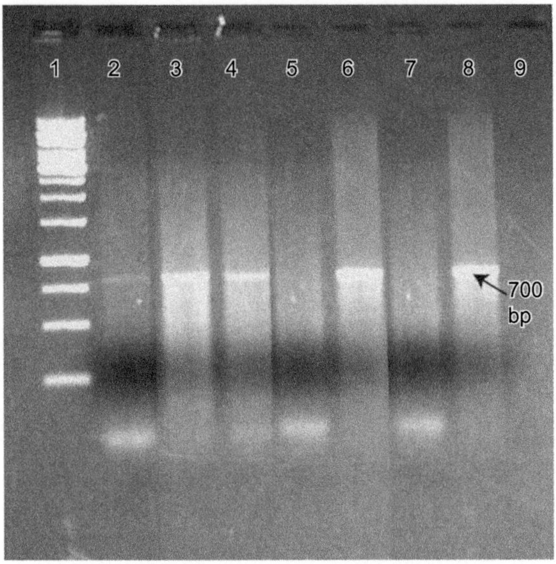

FIG. 8.41 Amplification of the *npt*II gene as determined by RT-PCR. Note the *npt*II gene-specific band (750 bp) was observed in four lanes (Lanes: 2, 3, 4, and 6), whereas there was no amplification in negative control (Lane: 7). Here, Lane 8 shows amplification of plasmid vector (positive control), and Lane: 9 refers to water control.

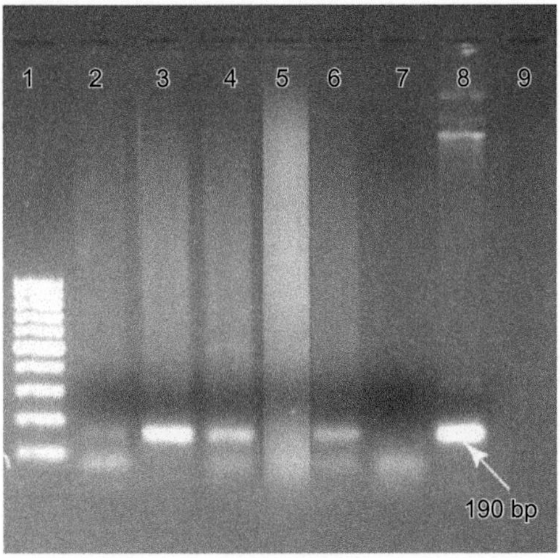

FIG. 8.42 Amplification of *AFP* gene as determined by RT-PCR. Note the *AFP* gene's specific band was observed in four lanes (Lanes: 2, 3, 4, and 6), whereas there was no amplification in negative control (Lane: 7). Here, Lane 8 shows amplification of the plasmid vector (positive control), and Lane: 9 refers to water control.

carried out using Northern hybridization to discover the expression of inserted transgenes in the RNA level. Among the five PCR positive transformed lines, three lines showed the expression of *npt*II genes.

8.10. CONCLUSIONS

The in-vitro regeneration system for the microsperma varieties of lentils has been optimized, and this regeneration protocol was found compatible with *Agrobacterium*-mediated genetic transformation. One notable finding was the development of in-vitro flowering and subsequent in-vitro seed formation from the in-vitro regenerated shoots of lentils to overcome the difficulties in obtaining effective root morphogenesis. The procedure of in-vitro flower and pod formation described here can be used for the successful recovery of in-vitro regenerated plantlets of lentils as well. This technique was also found to be effective toward the development of a transformation protocol, particularly for the development of transgenic progenies. Finally, it has been possible to develop transgenic lentil plantlets using both marker and antifungal genes which were confirmed through molecular characterization based on PCR and RT-PCR analysis, as well as Southern and Northern blotting techniques. Available literature indicated that this may be the pioneering report on the successful development of transgenic microsperma group of lentil plants, although the frequency of transformation using both marker and antifungal genes was rather low, which needs to be addressed in the future investigations on lentil genetic transformation. However, as an emerging technology, the *Agrobacterium*-mediated genetic transformation protocol described here can be utilized for the genetic improvement of lentils toward the development of various stress tolerance.

ACKNOWLEDGEMENTS

This study was supported by the United States Department of Agriculture (USDA) through a research grant to Rakha Hari Sarker. The authors would like to express their sincere thanks to Professor Dr. Hans-Jörg Jacobsen of the Department of Molecular Genetics, Leibniz University of Hannover, Germany, and to Dr. V. S. Reddy of International Centre for Genetic Engineering and Biotechnology (ICGEB), New Delhi, India, for providing technical cooperation. Seeds of lentil varieties used in this study were obtained from Bangladesh Agricultural Research Institute (BARI), Joydebpur, Gazipur, Bangladesh. The authors are also grateful to Dr. Tahmina Islam of the Department of Botany, University of Dhaka for her kind suggestions during the preparation of this manuscript.

REFERENCES

Akcay, U.C., Mahmoudian, M., Kamci, H., Yucel, M., Oktem, H.A., 2009. *Agrobacterium tumefaciens*-mediated genetic transformation of a recalcitrant grain legume, lentil (*Lens culinaris* Medik). Plant Cell Rep. 28, 407–417.

Amutha, S., Ganapathi, A., Muruganantham, M., 2003. *In vitro* organogenesis and plant formation in *Vigna radiata* (L.) Wilczek. Plant Cell Tissue Organ Cult. 72, 203–207.

Amutha, R., Jawahar, M., Ravi, P.S., 2008. Plant regeneration and *in vitro* flowering from shoot tip of *Basilicum polystachyon* (L.) Moench—an important medicinal plant. J. Agric. Sci. Technol. 4, 117–123.

Aragão, F.J.L., Vianna, G.R., Albino, M.M.C., Rech, E.L., 2002. Transgenic dry bean tolerant to the herbicide fluorinate ammonium. Crop Sci. 42, 1298–1302.

Asao, H., Nishizawa, Y., Arai, S., Sato, T., Hirai, M., Yoshida, K., Shinmyo, A., Hibi, T., 1997. Enhanced resistance against a fungal pathogen *Sphaerotheca humuli* in transgenic strawberry expressing a rice *chitinase* gene. Plant Biotechnol. J. 14, 145–149.

Atkins, C.A., Smith, P.M.C., 1997. Genetic transformation and regeneration of legumes. In: Legocki, A., Bothe, H., Piihler, A. (Eds.), Biological Fixation of Nitrogen for Ecology and Sustainable Agriculture. Springer, Berlin Heidelberg, New York, pp. 283–304.

Bahl, P.N., Lal, S., Sharma, B.M., 1993. An overview of the production and problems in Southeast Asia. In: Erskine, W., Saxena, M.C. (Eds.), Lentil in South Asia: Proceedings of the Seminar on Lentils in South Asia. ICARDA, Aleppo, Syria, pp. 1–10.

Bajaj, Y.P.S., Dhanju, M.S., 1979. Regeneration of plants from apical meristem tips of some legumes. Curr. Sci. 84 (20), 906–907.

Battraw, M.J., Hall, T.C., 1990. Histochemical analysis of CAMV 35S promoter *betaglucoronidase* gene expression in transgenic rice plants. Plant Mol. Biol. 15, 527–538.

Bayrac, A.T., 2004. Optimization of a Regeneration and Transformation System for Lentil (*Lens culinaris* Medik., cv. Sultan-I) Cotyledonary Petioles and Epicotyls (M.Sc. thesis). Middle East Technical University, Ankara, Turkey.

Bolar, J.P., Norelli, J.L., Wong, W.W., Hayes, C.K., Harman, G.E., Aldwinckle, H.S., 2000. Expression of endochitinase from *Trichoderma harzianum* in transgenic apple increases resistance to apple scab and reduces vigor. Phytopathology 90, 72–77.

Bressani, R., 1973. Legumes in Human Diet and How They Might be Improved. PAG of United Nations System15–42.

Bernier, G., 1988. The control of floral evocation and morphogenesis. Annu. Rev. Plant Physiol. Plant Mol. Biol. 39, 175–219.

Birch, R.G., 1997. Plant transformation: problems and strategies for practical application. Annu. Rev. Plant Physiol. Plant Mol. Biol. 48, 297–326.

Celikkol, U.A., Mahmoudian, M., Kamci, H., Yucel, M., Oktem, H.A., 2009. *Agrobacterium* tumefaciens-mediated genetic transformation of a recalcitrant grain legume, lentil (*Lens culinaris* Medik). Plant Cell Rep. 28, 407–417.

Chowrira, G.M., Akella, V., Lurquin, P.F., 1995. Electroporation mediated gene transfer into intact nodal meristems in planta. Mol. Biotechnol. 3, 17–23.

Chowrira, G.M., Akella, V., Feurst, P.E., Lurquin, P.F., 1996. Transgenic grain legumes obtained by in planta electoporation mediated gene transfer. Mol. Biotechnol. 5, 85–96.

Das, S.K., Shethi, K.J., Hoque, M.I., Sarker, R.H., 2012. *Agrobacterium*-mediated genetic transformation in lentil (*Lens culinaris* Medik.) followed by *In vitro* flowering and seed formation. Plant Tissue Cult. Biotechno. 22 (1), 13–26.

Dogan, D., Khawar, K.M., Ozcan, S., 2005. *Agrobacterium*-mediated tumor and hairy root formation from different explants of lentils derived from young seedlings. Int. J. Agric. Biol. 7, 1019–1025.

Doyle, J.J., Doyle, J.L., 1990. Isolation of plant DNA from fresh tissue. Focus 12, 13–15.

Eapen, S., Goerge, L., 1994. *Agrobacterium tumefaciens* mediated gene transfer in peanut (*Arachis hypogea* L.). Plant Cell Rep. 13, 582–686.

Erskine, W., 1997. Lessons for breeders from land races of lentil. Euphytica 93, 107–112.

Erskine, W., Sarker, A., 1997. Lentil: the Bangladesh breakthrough. Caravan 6, 8–9.

Fontana, G.S., Santini, L., Caretto, S., Frugis, G., Mariotti, D., 1993. Genetic transformation in the grain legume of *Cicer arietinum* L. Plant Cell Rep. 21, 194–198.

Fratini, R., Ruitz, M.L., 2003. Comparative study of different cytokinins in the induction of morphogenesis in lentil (*Lens culinaris* Medik.). In Vitro Cell. Dev. Biol. Plant 38, 46–51.

Fukamizo, T., Sasaki, C., Schelp, E., Bortone, K., Robertus, J.D., 2001. Kinetic properties of chitinase-1 from the fungal pathogen *Coccidiodes immitis*. Biochemistry 40, 2448–2454.

Gao, A., Hakimi, S., Mittanck, C., Wu, Y., Woerner, B., Stark, D., Shah, D., Liang, J., Rommens, C., 2000. Fungal pathogen protection in potato by expression of a plant defensin peptide. Nat. Biotechnol. 18, 1307–1310.

Gardner, R.C., 1993. Gene transformation into tropical and subtropical crops. Sci. Horticult. 55, 65–82.

Gheysen, G., Angenon, G., Van Montagu, M., 1998. Agrobacterium-mediated plant transformation: a scientifically intriguing story with significant application. In: Lindsey, K. (Ed.), Transgenic Plant Research. Harwood Academic, Amsterdam, pp. 1–33.

Gould, J.H., Smith, R.H., 1999. A non-distructive assay of GUS in the media of plant tissue cultures. Plant Mol. Biol. Rep. 7, 209–216.

Gowda, C.L., Kaul, A.K., 1982. 'Lentil' in Pulses in Bangladesh. Joint publication of Bangladesh Agricultural Research Institute and FAO51–106.

Gulati, A., Schryer, P., McHughen, A., 2001. Regeneration and micrografting of lentil shoots. In Vitro Cell. Dev. Biol. Plant 37 (6), 798–802.

Gulati, A., Schryer, P., McHugen, A., 2002. Production of fertile transgenic lentil (*Lens culinaris* Medik) plants using particle bombardment. In Vitro Cell. Dev. Biol. Plant 38, 316–324.

Halbach, T., Kiesecker, H., Jacobsen, H.J., DeKathen, A., 1998. Tissue culture and genetic engineering of lentil (*Lens culinaris* Medik.). In: 3rd European Conference on Grain Legumes, Valladolid. pp. 376.

Hassan, F., Hoque, M.I., Kiesecker, H., Jacobsen, H., 2007. Transient GUS expression in decapitated lentil embryos. Plant Tissue Cult. Biotechnol. 17 (1), 97–102.

Hellens, R.P., Edwards, A., Leyland, N.R., Bean, S., Mullineaux, P.M., 2000. pGreen: a versatile and flexible binary Ti vector for *Agrobacterium*-mediated plant transformation. Plant Mol. Biol. 42, 819–832.

Herrea-Esterella, L., Depicker, A., Montagu, M., Schell, J., 1983. Expression of chimeric genes transferred into plant cells using a Ti-plasmid derived vector. Nature 303, 209.

Heslop-Harrison, J., Heslop-Harrison, Y., 1970. Evaluation of pollen viability by enzymatically induced fluorescence: intracellular hydrolysis of fluorescein diacetate. Stain. Technol. 45, 115–120.

Hinchee, M.A.W., Connor-Ward, D.V., Newell, C.A., McDonell, R.E., Sato, S.J., Gasser, C.S., Fischhoff, D.A., Re, D.R., Fraley, R.T., Horsch, R.B., 1988. Production of transgenic soyabean plants using *Agrobacterium*-mediated DNA transfer. Biotechnology 6, 915–922.

Hood, E.E., Gelvin, S.B., Melchers, L.S., Hoekema, A., 1993. New *Agrobacterium* helper plasmids for gene transfer to plants. Transgenic Res. 2, 208–218.

Hooykaas, P.J.J., 1988. Agrobacterium molecular genetics. Plant Mol. Biol. Manual 4, 1–13.

Hoque, M.I., Hassan, F., Sarker, R.H., Kisecker, H., Jacobsen, H.J., 2004. Lentil improvement through biotechnology. In: Islam, A.S. (Ed.), In Vitro Culture, Transformation and Molecular Markers for Crop Improvement. Science Publishers, Incorporation, Enfield, USA, pp. 175–192.

Hulse, J.H., 1994. Nature, composition and utilization of food legumes. In: Muehlbauer, F.J., Kaiser, W.J. (Eds.), Expanding the Production and Use of Cool Season Food Legumes. Kluwar Academic Publishers, The Netherlands, pp. 235–261.

ICARDA, 2000. Lentil is Good for Food and Good for Environment [Brochure]. Aleppo-Syria.

Ikea, J., Ingrlbrecht, I., Uwaifo, A., Thottrppilly, G., 2003. Stable gene transformation in cowpea (*Vigna unguiculata* L. walp.) using particle gun method. Afr. J. Biotechnol. 2 (8), 211–218.

Islam, A., 2006. Fungus resistant transgenic plants: strategies, progress and lessons learnt. Plant Tissue Cult. Biotechnol. 16 (2), 117–138.

Jackson, J.A., Hobbs, S.L.A., 1990. Rapid multiple shoot production from cotyledonary node explants of pea (*Pisum sativum* L.). In vitro Cell. Dev. Biol. 26, 835–838.

Jacobson, H.J., 1992. In: Biotechnology applied to grain legumes—current state and prospects. Proceedings of 1st European Conference on Grain Legumes, Angers, France. pp. 99–103.

Jaiwal, P.K., Kumari, R., Ignacimuthu, S., Potrykus, I., Curistos, S., 2001. *Agrobacterium tumefaciens* genetic transformation of mungbean (*Vigna radiata* L. wilczek.) a recalcitrant grain legume. Plant Sci. Shannon 161, 239–247.

Jayanand, B., Sudarsanam, G., Sharma, K.K., 2003. An efficient protocol for the regeneration of whole plants of chickpea (*Cicer arietinum* L.) by using axillary meristem explants derived from *in vitro*-germinated seedlings. In Vitro Cell. Dev. Biol. Plant 39, 171–179.

Jefferson, R.A., Kavanagh, T.A., Bevan, M.W., 1987. GUS fusion: beta-glucoronidase as a sensitive and versatile gene fusion marker in higher plants. EMBO J. 6, 3901–3907.

Jumin, H.B., Ahmed, M., 1999. High-frequency *in vitro* flowering of *Murraya paniculata* (L.) Jack. Plant Cell Rep. 18, 764–768.

Kanchanapoom, K., Posayapisit, N., Kanchanapoom, K., 2009. *In Vitro* flowering from cultured nodal explants of rose (*Rosa hybrida* L.). Notulae Botanicae Horti Agrobotanici Cluj-Napoca 37, 261–263.

Kaneda, Y., Tabei, Y., Nishimura, S., Harada, K., Akihama, T., Kitamura, K., 1997. Combination of thidiazuron and basal media with low salt concentration increase the frequency of shoot organogenesis in soybeans [*Glycine max* (L.) Merr.]. Plant Cell Rep. 17, 8–12.

Kar, S., Johanson, T.M., Nayak, P., Sen, S.K., 1996. Efficient transgenic plant regeneration through *Agrobacterium*-mediated transformation of chickpea. Plant Cell Rep. 16, 32–37.

Kartha, K.K., Pahl, K., Leung, N.L., Mroginski, L.A., 1981. Plant regeneration from meristems of grain legumes soybean cowpea, peanut, chickpea and bean. Can. J. Bot. 59, 167–179.

Katerji, N., Van Hoorn, J.W., Hamdy, A., Mastrorilli, M., 2003. Salinity effect on crop development and yield analysis of salt tolerance according to several classification methods. Agric. Water Manag. 62, 37–66.

Khan, M.R.I., Tabe, L.M., Heath, L.C., Spencer, D., Higgins, T.J.V., 1994. *Agrobacterium*-mediated genetic transformation of subterranean clover (*Trifolium subterraneum* L.). Plant Physiol. 105 (2), 81–88.

Khanam, R., Sarker, R.H., Hoque, M.I., Haque, M.M., 1995. *In vitro* root morphogenesis in lentil (*Lens culinaris* Medik). Plant Tissue Cult. 5, 35–41.

Khatib, F., Makris, A., Yamaguchi-Shinozaki, K., Kumar, S., Sarker, A., Erskine, W., Baum, M., 2011. Expression of the *DREB1A* gene in lentil (*Lens culinaris* Medik. Subsp. *culinaris*) transformed with the *Agrobacterium* system. Crop Pasture Sci. 62, 488–495.

Khawar, K., Ozcan, S., 2002. *In vitro* induction of crown galls by *Agrobacterium tumefaciens* super virulent strain $A_2$81 (pTIBo542) in lentil (*Lens culinaris* Medik.). Turkey J. Bot. 26, 165–170.

Khawar, K.M., Sancak, C., Uranbey, S., Ozcan, S., 2003. Effect of Thidiazuron on shoot regeneration from different explants of lentil (*Lens culinaris* Medik.) via organogenesis. Turkish J. Bot. 28, 421–426.

Khazaei, H., Caron, C.T., Fedoruk, M., Diapari, M., Vandenberg, A., Coyne, C.J., McGee, R., Bett, K.E., 2016. Genetic diversity of cultivated lentil (*Lens culinaris* Medik.) and its relation to the world's agro-ecological zones. Front. Plant Sci. 7, 1093.

Kho, Y.O., Baer, J., 1968. Observing pollen tubes by means of fluorescence. Euphytica 17, 198–302.

Krishna, G., Singh, B.K., Kim, E.K., Morya, V.K., Ramteke, P.W., 2015. Progress in genetic engineering of peanut (*Arachis hypogaea* L.)—a review. Plant Biotechnol. J. 13 (2), 147–162.

Krishnamurthy, K.V., Suhasini, K., Sagare, A.P., Meixner, M., De Kathen, A., Pickardt, T., Schieder, O., 2000. *Agrobacterium*-mediated genetic transformation of chickpea (*Cicer arietinum* L.) embryo axes. Plant Cell Rep. 19, 235–240.

Ladizinsky, G., 1979. The origin of lentil and its wild gene pool. Euphytica 28, 179–187.

Ladizinsky, G., Cohen, D., Muehlbauer, F.J., 1985. Hybridization in the genus *Lens* by means of embryo culture. Theor. Appl. Genet. 70, 97–101.

Larkin, P.J., Scowcroft, W.R., 1981. Somaclonal variation—a novel source of variability from cell cultures for plant improvement. Theor. Appl. Genet. 60, 197–214.

Lorito, M., Woo, S.L., Fernandez, I.G., Colucci, G., Harman, G.E., Pinto-Tora, J.A., Fillipone, E., Muccifora, S., Lawrence, C.B., Zoina, A., Tuzun, S., Scala, F., 1998. Genes from mycoparasitic fungi as a source for improving plant resistance to fungal pathogens. Proc. Natl. Acad. Sci. U. S. A. 95, 7860–7865.

Lurquin, P.F., Cai, Z., Stiff, C.F., Fuerst, E.P., 1998. Half embryo co-cultivation technique for estimating the susceptibility of pea (*Pisum sativum* L.) and lentil (*Lens culinaris* Medik.) cultivars to *Agrobacterium tumefaciens*. Mol. Biotechnol. 9, 175–179.

Maccarrone, M., Veldink, G.A., AgrÒ, A.F., Vliegenthart, J.F.G., 1995. Lentil root protoplast: a transient expression system suitable coelectroporation of monoclonal antibodies and plasmid molecules. Biochem. Biophys. Acta 1243, 136–142.

Mahmoudian, M., Yücel, M., Öktem, H.A., 2002. Transformation of lentil (*Lensculinaris* M.) cotyledonary nodes by vaccum infiltration of *Agrobacterium tumefacien*. Plant Mol. Biol. Report. 20, 251–257.

Malik, M.A., Rashid, A., 1989. Induction of multiple-shoots from coteledonary node of grain legumes: pea and lentil. Biol. Plant. 31, 230–232.

Malik, K.A., Saxena, P.K., 1992. Thiadiazuron induces high frequency shoot regeneration in intact seedlings of pea (*Pisum sativum* L.), chickpea (*Cicer arietinum*), lentil (*Lens culinaris*). Aust. J. Plant Physiol. 19, 731–740.

Mandal, A.P., Maiti, A., Elanchezhian, R., 2000. *In vitro* flowering in maize (*Zea mays* L.). Asia-Pac. J. Mol. Biol. Biotechnol. 8, 81–83.

Mansur, E.A., Lacorte, C., DeFreitus, V.G., De Olivierea, D.E., Timmerman, B., Cordeiro, A.R., 1993. Regulation of transformation efficiency of peanut (*Arachis hypogea* L.) explants by *Agrobacterium tumefaciens*. Plant Sci. 89, 93–99.

McCourt, P., 1999. Genetic analysis of hormone signaling. Annu. Rev. Plant Physiol. Plant Mol. Biol. 50, 219–243.

McHughen, A., Jordan, M., Feist, G., 1989. A preculture period prior to *Agrobacterium* inoculation increases production of transgenic plants. J. Plant Physiol. 135, 245–248.

McKently, A.H., Moore, G.H., Doostdar, H., Niedz, R.P., 1995. *Agrobacterium*-mediated genetic transformation of peanut (*Arachis hypogea* L.) embryo axes and the development of transgenic plants. Plant Cell Rep. 14, 699–703.

Meurer, C.A., Dinkins, R.D., Collins, G.B., 1998. Factors affecting soyabean cotyledonary node transformation. Plant Cell Rep. 18, 180–186.

Monti, L., Biddle, A.J., Moreno, M.T., Plancquarte, P., 1994. 'Biotic and abiotic stresses of pulse crops in Europe. In: Muehlbauer, F.J., Kaiser, K.J. (Eds.), Expanding the Production and Use of Cool season Food Legumes. Kluwer Academic Publishers, Dordrecht, The Netherlands, pp. 204–218.

Moss, G.I., 1969. Influence of temperature and photoperiod of flower induction and inflorescence development in sweet orange (*Citrus sinensis* L. Osbeck). J. Horticult. Sci. 44, 311–320.

Mroginski, L.A., Kartha, K.K., 1984. Tissue culture of legumes for crop improvement. In: Janick, J. (Ed.), Plant Breeding Revision. 2. The AVI Publishing Company Incorporation, Westport, Cannecticut, pp. 215–264.

Muehlbauer, F.J., McPhee, K.E., 2002. In: Brouwer, J.B. (Ed.), Future of North American lentil production. Proceedings of Lentil Focus 2002 Meeting. Horsham, Victoria, Australia.

Murashige, T., Skoog, F., 1962. A revised medium for rapid growth and bioassays with tobacco tissue culture. Plant Physiol. 15, 473–497.

Nadgauda, R.S., Parasharami, V.A., Mascarenhas, A.F., 1990. Precocious flowering and seedling behaviour in tissue-cultured bamboos. Nature 344, 335–336.

Nagasaki, H., Yamamoto, K., Shomura, A., Koga-ban, Y., Takasuga, A., Yano, M., Minobe, Y., Sasaki, T., 1997. Rice class III *chitinase* homologues isolated by random cloning of rice cDNAs. DNA Res. 4, 379–385.

Neuhaus, JM, Ahl-Goy, P. Hinz, U. Flores, S & Meins, FJr 1991, 'High-level expression of a tobacco *chitinase* gene in *Nicotiana sylvestris*. Susceptibility of transgenic plants to *Cercospora nicotianae* infection', Plant Mol. Biol., vol. 16, no. 1, pp. 141–151.

Nirala, N.K., Das, D.K., Srivastava, P.S., Sopory, S.K., Upadhyaya, K.C., 2010. Expression of a rice *chitinase* gene enhances antifungal potential in transgenic grapevine (*Vitis vinifera* L.). Vitis 49 (4), 181–187.

Nisbet, G.S., Webb, K.J., 1990. Legumes and Oilseed Crops. In: Bajaj, Y.P.S. (Ed.), Biotechnology in Agriculture and Forestry. vol. 10(1). Springer-Verlag, Berlin, pp. 38–48.

Öktem, H.A., Mahmoudian, M., Eyidodan, F., Yucel, M., 1999. GUS gene delivary and expression in lentil cotyledonary nodes using particle bombardment. Lens Newsl. 26 (1 and 2), 3–6.

Oram, P.A., Agacoli, M., 1994. Current status and future trends in supply and demand of cool season food legumes. In: Muehlbauer, F.J., Kaiser, W.J. (Eds.), Expanding the Production and Use of Cool Season Food Legumes. Kluwer Academic Publisher, The Netherlands, pp. 3–49.

Ottaviani, M.P., Smits, T., Hanish, C.H., 1993. Differential methylation and expression of the *β-glucuronidase* and *neomycin phosphotransferase* genes in transgenic plants of potato cv. Bintje. Plant Sci. 88, 73–81.

Pezzoti, M., Pupilli, F., Damiani, F., Arcioni, S., 1991. Transformation of *Medicago sativa* L. using a Ti-plasmid derived vector. Plant Breed. 106, 39–46.

Polanco, M.C., Pelaez, M.I., Ruiz, M.L., 1988. Factor affecting callus and shoot formation from *in vitro* cultures of *Lens culinaris* Medik. Plant Cell Tissue Organ Cult. 15 (2), 175–182.

Polanco, M.C., Ruiz, M.L., 1997. Effect of benzyleaminopurine on *in vitro* and *in vivo* root development in lentil (*Lens culinaris* Medik.). Plant Cell Rep. 17, 22–26.

Polanco, M.C., Ruiz, M.L., 2001. Factors that affect plant regeneration from *in vitro* culture of immature seeds in four lentil (*Lens culinaris* Medik.) cultivars. Plant Cell Tissue Organ Cult. 66, 133–139.

Polowick, P.L., Quandat, J., Mahon, J.D., 2000. The ability of pea transformation technology to transfer genes into peas adapted to western Canadian growing conditions. Plant Sci. 153, 161–170.

Rahman, M.L., Bakr, M.A., 1998. In: Singh, K.B., Rahman, M.M. (Eds.), Current status of lentil diseases and future needs. Proceedings Workshop on Diseases Research Breeding in Pulses. BARI, Joydebpur, Gazipur, Bangladesh, pp. 23–31.

Rajasekaran, K., Cary, J.W., Jacks, T.J., Stromberg, K.D., Cleveland, T.E., 2000. Inhibition of fungal growth in planta and *in vitro* by transgenic tobacco expressing a bacterial nonheme chloroperoxidase gene. Plant Cell Rep. 19, 333–338.

Saini, R., Jaiwal, S., Jaiwal, P.K., 2003. Stable genetic transformation of *Vigna mungo* L. Hepper via *Agrobacterium tumefaciens*. Plant Cell Rep. 21, 851–859.

Sambrook, J., Fritsch, E.F., Maniatis, T., 1989. Molecular Cloning, a Laboratory Manual, second ed. Cold Spring Harbor Laboratory Press, Cold Spring Harbor, New York.

Sarker, R.H., Biswas, A., 2002. *In vitro* plantlet regeneration and *Agrobacterium*-mediated genetic transformation of wheat (*Triticum aestivum* L.). Plant Tissue Cult. 12 (2), 155–165.

Sarker, R.H., Das, S.K., Hoque, M.I., 2012. *In vitro* flowering and seed formation in lentil (*Lens culinaris* Medik.). In vitro Cell. Dev. Biol. Plant 48, 446–452.

Sarker, R.H., Mustafa, B.M., Biswas, A., Mahbub, S., Hoque, M.I., 2003a. *Agrobacterium*-mediatedtransformation of lentil (*Lens culinaris* Medik.). Plant Tissue Cult. 13 (1), 1–12.

Sarker, R.H., Mustafa, B.M., Biswas, A., Mahbub, S., Nahar, M., Hashem, R., Hoque, M.I., 2003b. *In vitro* regeneration in lentil (*Lens culinaris* Medik.). Plant Tissue Cult. 13 (2), 155–163.

Saxena, M.C., 1993. The challenge of developing biotic and abiotic stress resistance in cool-season food legumes. In: Singh, K.B., Saxena, M.C. (Eds.), Breeding for Stress Tolerance in Cool Season Food Legumes. Wiley, Chichester, United Kingdom, pp. 3–14.

Saxena, P.K., King, J., 1987. Morphogenesis in lentil: plant regeneration from callus culture of *Lens culinaris* Medik. via somatic embryogenesis. Plant Sci. 52, 223–227.

Schroeder, H.E., Schotz, A.H., Wardely-Richardson, T., Spencer, D., Higgins, T.J.V., 1993. Transformation and regeneration of two cultivars of pea (*Pisum sativum* L.). Plant Physiol. 101, 751–757.

Seraj, Z.I., Sarker, A.B., Islam, A.S., 1992. Plant regeneration in a jute species (*C. capsularis*) and its possible relationship with glyoxalase-I. Plant Cell Rep. 12 (1), 29–33.

Sevimoy, C.S., Khawar, K.M., Çöçü, S., Özcan, S., 2005. Somatic embryogenesis in white clover (*Trifolium repens* L.). Period. Biol. 107 (1), 101–105.

Sheeja, T.E., Mandal, A.B., 2003. *In vitro* flowering and fruiting in tomato (*Lycopersicon esculentum* Mill.). Asia Pac. J. Mol. Biol. Biotechnol. 11, 37–42.

Shimelis, H., Laing, M., 2012. Timelines in conventional crop improvement: pre-breeding and breeding procedures. Aust. J Crop Sci. 6 (11), 1542–1549.

Singh, R.K., Raghuvanshi, S.S., 1989. Plantlet regeneration from nodal segment and shoot tip derived explant of lentil. Lens Newsl. 16 (1), 33–35.

Singh, M., Jaiswal, U., Jaiswal, V.S., 2000. Thidiazuron induced *in vitro* flowering in *Dendroclamus strictus* Nees. Curr. Sci. 79, 1529–1530.

Subhadra, Vashishth, R.K., Chowdhury, J.B., Singh, M., Sareen, P.K., 1998. Multiple shoots from cotyledonary node explants of non noduling genotype (ICC435M) of chickpea, *Cicer arietinum* L. Indian J. Exp. Biol. 36, 1276–1279.

Surekha, S.H., Beena, M.R., Arundhati, A., Singh, P.K., Tuli, A., Gupta, A.D., Kirti, P.B., 2005. *Agrobacterium*-mediated genetic transformation of pigeon pea [*Cajanus cajan* (L.) Millsp.] using embryonal segments and development of transgenic plants for resistance against *Spodoptera*. Plant Sci. 169 (6), 1074–1080.

Tang, W., 2000. High-frequency plant regeneration *via* somatic embryogenesis and organogenesis and *in vitro* flowerng of regenerated plantlets in *Panax ginseng*. Plant Cell Rep. 19, 727–732.

Tewari-Singh, N., Sen, J., Kiesecker, H., Reddy, V.S., Jacobson, H.J., Mukherjee, S.G., 2004. Use an herbicide or lysine plus threonine for non antibiotic selection of transgenic chickpea. Plant Cell Rep. 22, 576–583.

Trieu, A.T., Burleigh, S.H., Kardailsky, I.V., Maldonado-Mendoza, I.E., Versaw, W.K., Blerplock, L.A., Shin, M., Chiou, T.J., Katageri, H., Derobre, G.R., Weigel, D., Harrison, M.J., 2000. Transformation of *Medicago truncatula* via infilteration of seedlings or flowering plants with *Agrobacterium*. Plant J. 22, 531–541.

Tztzikas, E.N., Bergervoet, M., Reamakers, K., Vincken, J.P., Van Lammeren, A., Visser, R.G.F., 2004. Regeneration of pea (*Pisum sativum* L.) by a cyclic organogenic system. Plant Cell Rep. 23, 453–460.

Vandana, A.K., Kumar, A., Kumar, J., 1995. *In vitro* flowering and pod formation in cauliflower (*B. oleracea* var. *botrytis*). Curr. Sci. 69, 543–545.

Van den Bergh, K.P.B., Rouge, P., Proost, P., Coosemans, J., Krouglova, T., Engelborghs, Y., Peumanas, W.J., Van Damme, E.J.M., 2004. Synergatic antifungal activity of two chitin—binding proteins from spindle tree (*Euonymus europaeus* L.). Planta 219, 221–232.

Wang, S., Tang, L., Chen, F., 2001. In vitro flowering of bitter melon. Plant Cell Rep. 20, 393–397.

Warkentin, T.D., McHughen, A., 1991. Crown gall formation of lentil (*Lens culinaris* Medik.) with virulent strains of *Agrobacterium tumefaciens*. Plant Cell Rep. 10, 489–493.

Warkentin, T.D., McHughen, A., 1992. *Agrobacterium tumefaciens*-mediated *beta-glucoronidase* (*GUS*) gene expression in lentil (*Lens culinaris* Medik.) tissues. Plant Cell Rep. 11, 274–278.

Warkentin, T.D., McHughen, A., 1993. Regeneration from lentil cotyledonary nodes and potential of this explant for transformation by *Agrobacterium tumefaciens*. Lens Newsl. 20 (1), 26–28.

Weller, J.T., Reid, J.B., Tayler, S.A., Murfet, I.C., 1997. The genetic control of flowering in pea. Trends Plant Sci. 2, 412–418.

Weigel, D., 1995. The genetics of flower development: from floral induction to ovule morphogenesis. Annu. Rev. Genet. 29, 19–39.

Wiese, W.B., Vorman, E., Krause, H.K., 1994. Structural organization and differential expression of three *stilbene synthase* genes located on a 13 kb grapevine DNA fragment. Plant Mol. Biol. 26, 667–677.

Williams, D.J., McHugen, A., 1986. Plant regeneration of legume *Lens culinaris* Medik (lentil) *in vitro*. Plant Cell Tissue Organ Cult. 7 (2), 149–153.

Ye, G., McNeil, D.L., Hill, G.D., 2002. Breeding for resistance to lentil *Ascochyta blight*. Plant Breed. 121, 185–191.

Ying, M., Dyer, W.E.B., Gergman, I.W., 1992. *Agrobacterium tumefaciens*-mediated transformation of safflower (*Carthamus tinctorius* L.) cv. Centennial. Plant Cell Rep. 11, 581–585.

Zohary, D., 1986. Lentil. In: Simmonds, N.M. (Ed.), Evolution of Crop Plants. Longman Scientific and Technical, Longman Group, UK Limited, pp. 339.

Chapter 9

The Lentil Economy in India

Nisha Varghese*, Atul Dogra†, Ashutosh Sarker†, Aden Aw Hassan‡
**School of Extension and Development Studies, Indira Gandhi National Open University, New Delhi, India, †International Center for Agricultural Research in the Dry Areas, South Asia & China Regional Program, New Delhi, India, ‡International Center for Agricultural Research in the Dry Area, ICARDA, Amman, Jordan*

9.1. INTRODUCTION

Pulses have been an integral part of the human diet for centuries. Pulses, being a rich source of proteins, fibers, minerals and vitamins, are important for food security, combating malnutrition, and poverty alleviation. Due to their high nutrient density, they are also used in fortifying other foodstuffs, especially cereals. Pulses are grown in both developed and developing countries. In developed countries, they are grown in large scale, using mechanized farming, and are generally used as animal feed. In developing countries, these are grown by resource-poor small and marginal farmers, for whom pulses contribute toward household food security and serve as cash crops for income generation.

Pulses are particularly important, given that the largest number of malnourished people in the world live in India. A large percentage of Indians suffer from nutrition problems, including protein deficiency. Given the levels of poverty and prevalence of vegetarianism, ameliorating the problem of protein deficiency suggests an important place for pulses (Roy et al., 2016). With India's large vegetarian population, pulses constitute the most common source of protein (the frequency of its consumption is higher than any other source of protein) contributing to more than 10% of protein intake. Around 89% of consumers have pulses at least once a week in India, while the corresponding number for fish or chicken/meat is only 35.4% (IIPS and Macro International, 2007). Animal source foods (ASF) are the most expensive source of protein in India, making pulses the affordable alternative. Pulses are not only important because of their nutritional value; they also enhance soil productivity and can improve the yield of other crops. With these benefits, pulses should have received adequate attention from growers, consumers, technologists, and policymakers, alike. Unfortunately, that has not been the case for pulses. Unlike cereals, the per capita urban-to-rural consumption gap for pulses has been widening over the years due to the more rapid decline in pulse consumption in rural areas (Ali and Gupta, 2012). Food-insecure

Lentils. https://doi.org/10.1016/B978-0-12-813522-8.00009-1

farmers' reliance upon cereal-dominated cropping systems presents a challenge to the long-term sustainability of cropping systems in developing countries (Kerr et al., 2007). India has persistently faced a deficit, and a sustainable part of domestic demand has been met by imports for over one and a half decades now. Recognizing the importance and potential of pulses for agricultural systems, nutrition, and food security, in 2013, the United Nations General Assembly declared 2016 the International Year of Pulses (FAO, 2016) (Table 9.1).

9.2. GLOBAL PULSE ECONOMY: AN OVERVIEW

Since the 1960s, the production of pulses has made steady advances across the world. Asia and Africa account for around 70% of world pulse production (Fig. 9.1). In developing countries, pulses are grown for subsistence and are often grown as crops secondary to staples. But in developed countries, pulses are grown on an industrial scale. For the past few decades, India has continuously been the largest producer of pulses. In Asia, Myanmar increased its pulse production by more than 20-fold since the green revolution, and today, it is the world's third largest producer of pulses. On the other hand, pulse production in China has decreased by more than half during the same period, indicating a shift in consumer preferences in favor of animal-based proteins.

Area-based growth rates for almost all the pulses (except chickpea) declined from 1981 to 2000. The total areas sown in cowpeas and pigeon peas have shown tremendous growth in the current decade. The growth rate in area for lentils since 2011 has also considerably increased compared to the previous decades. Among pulses, the production of dry beans and chickpeas has been growing steadily over the decades. Since 1991, lentil production has grown; cowpea and dry pea production has fluctuated over the years, and pigeon pea production has grown steadily over the decades. Consumption of pulses has been declining in both developed and developing countries, probably due to consumers' dietary preference for dairy and animal protein over plant protein. Per capita consumption of pulses is not widely measured. The total availability of pulses for food across countries ranges from approximately 2 kg per capita per year in high-income countries to 10–15 kg in middle-income countries, mostly in Asia and Latin America (McDermott and Wyatt, 2017). The growth rate in productivity for almost all the pulses (except beans and chickpeas) has been very low. In cowpeas and pigeon peas, the growth rate has been negative in the current decade (Table 9.2).

There are multiple reasons for stagnating pulse production. Pulses are cultivated in unirrigated areas, to a great extent, on what is generally poor quality acreage. There has been no major technological breakthrough yet which produced high-yielding varieties of pulses. The few available improved varieties not widely adaptable, and they disease-prone. Certified seeds can be difficult to obtain, and when it is an option, farmers with access to irrigation prefer to plant more lucrative crops (Sathe and Agarwal, 2004). The lack of an assured market

TABLE 9.1 Nutritive Value of Different Pulses

Pulses	Energy (Kcals)	Protein (g)	Fat (g)	Mineral (g)	Carbohydrates (g)	Fiber (g)	Calcium (mg)	Phosphorous (mg)	Iron (mg)
Gram	360	17	5	3	61	4	202	312	5
Cowpeas	323	24	1	3	54	3	77	414	9
Green grams	334	24	1	3	57	4	124	326	4
Horse grams	321	22	0	3	57	5	287	311	7
Lentils	343	25	1	2	59	1	69	293	7
Peas	315	20	1	2	56	4	75	298	7
Kidney beans	346	23	1	3	61	5	260	410	5

Source: Gopalan, C., Rama Sastri, B.V., Balasubramanyan, S.C., Narasinga Rao, B.S., Deosthale, Y.G., Pant, K.C., 2004. Nutritive Value of Indian Foods. National Institute of Nutrition, ICMR, Hyderabad.

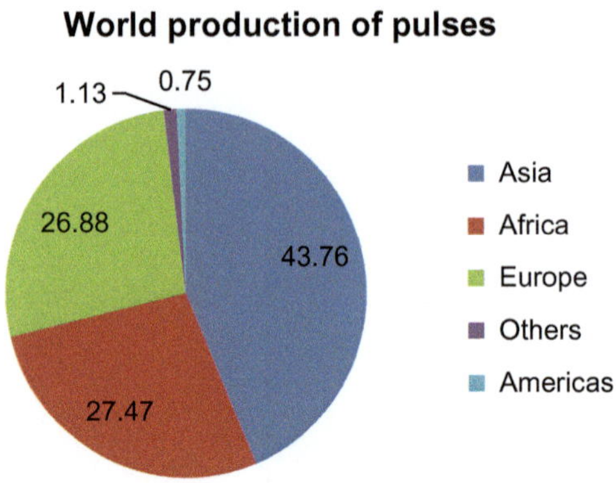

FIG. 9.1 Region regional world production of pulses 2014.

is another factor in pulses' poor performance. Compared to the market for cereals in many parts of the country, the markets for pulses are thin and fragmented (Reddy, 2004).

Pulses have a volatile market. There has been a spur in the demand for pulses in the developing countries due to income growth. To meet this demand, some developed countries have increased their pulse production to exploit this burgeoning market. India remains both the highest producer and consumer of pulses. Some of the major pulse-importing countries are India, Pakistan, China, Egypt, the United States, Turkey, and Italy. The major pulse-exporting countries are Canada, Australia, the United States, China, the Russian Federation, and Argentina.

9.3. LENTILS: AN OVERVIEW

As a percentage of the total number of pulses, peas, broad beans, vetches, and pulses have decreased from the 1960s to the present decade, whereas cowpeas, dry beans, pigeon peas, and lentils have increased their proportion in the overall total. The share of lentils, in particular, has increased from 2.25% of the total in the 1960s to 6.33% in the current decade. Lentils were one of the first domesticated plant species; they are as old as einkorn, emmer, barley, and peas (Harlan, 1992). Lentil seeds come in a range of sizes and colors. They are lens-shaped (the word "lentil" is derived from the Latin word "lens"), generally from 2 to 9 mm in diameter, and vary from 15,600 to 100,000 seeds per pound (Sell, 1993). Though small in size, lentils are gaining importance in human diet due to their ease of preparation and high nutritional value. Lentils are the richest source of dietary protein among plants that are consumed without industrial processing (Sharma, 2009). Lentils are also a very good source of cholesterol-lowering fiber and are beneficial in managing blood-sugar-related disorders. Lentils are a good source of micronutrients like molybdenum, folate, copper, phosphorus, manganese, and iron. Usually, lentils are grown in the subtropics

TABLE 9.2 Growth Rates in Area, Production, and Yield of Pulses Since 1981

Year	Beans Dry	Chickpea	Cowpea	Lentils	Peas Dry	Pigeon Pea
Growth in Area of Pulses Since 1981–2014						
1981–1990	0.15	0.12	5.40	4.08	2.21	2.59
1991–2000	−0.96	1.03	2.78	1.37	−3.14	0.14
2001–2010	1.15	2.32	2.25	0.19	0.55	0.90
2011–2014	−0.19	1.67	6.24	2.87	1.22	6.33
Growth in Production of Pulses Since 1981–2014						
1981–1990	1.08	1.13	6.69	6.07	7.91	3.1
1991–2000	0.58	2.47	4.49	1.77	−2.48	1.57
2001–2010	2.19	4.41	5.71	2.94	−0.14	3.34
2011–2014	4.28	6.26	4.9	3.8	2.64	3.63
Growth in Yield of Pulses Since 1981–2014						
1981–1990	0.92	1.01	1.23	1.91	5.58	0.50
1991–2000	1.56	1.42	1.67	0.40	0.68	1.43
2001–2010	1.03	2.04	3.38	2.74	−0.69	2.42
2011–2014	4.48	4.51	−1.26	0.90	1.40	−2.54

Source: FAO, 2016. Food Outlook: Biannual Report on Global Food Markets. FAO, p. 139.

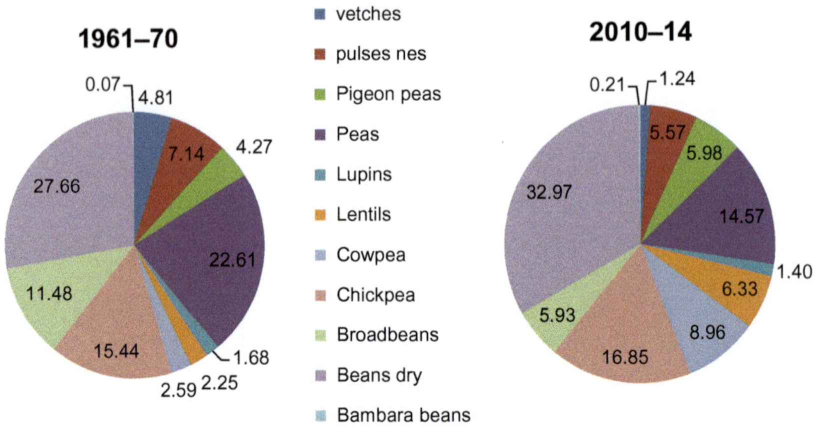

FIG. 9.2 Change in the Share of Different Pulses in the Total Pulse Basket During the 1960s and the Current Decade.

with winter rainfall, in warm temperate regions, and in the tropics and subtropics with summer rainfall, either during cool dry seasons or at high altitudes (Al-Issa, 2006) (Fig. 9.2).

Asia and the Americas contribute around 90% of the world's lentil-sown area and lentil production. In the Americas, lentils are mainly grown in Canada and the United States. India is the largest producer of lentils in Asia. Other lentil-producing countries include Turkey, Nepal, China, Bangladesh, Syria, and Iran. While around 61% of the world's area sown in lentils is in Asia, it produces only 44% of the world's lentil harvest. The Americas, on the other hand, have only around 30% of the world's lentil-sown area, contributing to around 45% of the world's lentil production. This shows wide variation in lentil yields between the developed and developing worlds. The average lentil yield in the Americas (1607 kg/ha) is more than double the average yield in Asia (778 kg/ha), which is the main reason for Asia's smaller share of world production (Fig. 9.3).

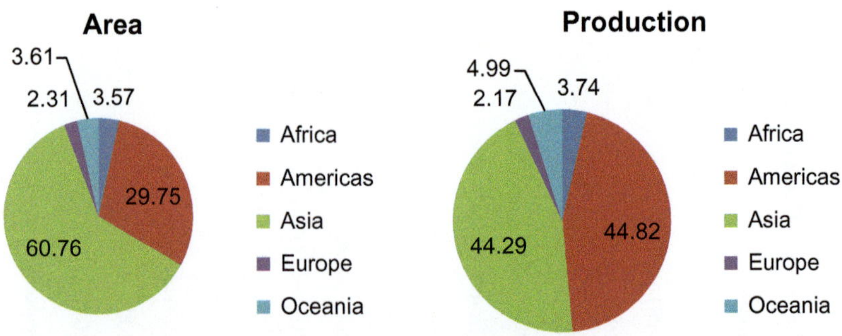

FIG. 9.3 Share of different regions in area and production of pulses, 2014.

On average, about 70% of all the world's lentil production is consumed in the countries where they are produced (Singh and Singh, 2014). The major exporters of lentils are Canada, Turkey, Australia, and the United States, and the leading importers are India, Pakistan, China, Egypt, the United States, Turkey, Italy, Algeria, the UK, and Sri Lanka.

9.4. INDIAN PULSE SCENARIO

Pulses occupy an important and unique place in the Indian food basket. Pulses are particularly important given that the largest number of malnourished people in the world live in India. A large percentage of Indians suffer from nutrition problems including protein deficiency (Roy et al., 2016). Not only is India the largest producer and consumer of pulses, it is also one of the highest pulse-importing countries. Pulses in India are grown on marginal lands under rain-fed conditions with very low input levels, resulting in low productivity. In 2015–16, India produced around 16.35 million tons of pulses. Madhya Pradesh, Maharashtra, Rajasthan, Uttar Pradesh, Karnataka, Andhra Pradesh, Gujarat, Tamil Nadu, Jharkhand, and Bihar are the major pulse producing states in India.

Because pulses are the major protein source for the large vegetarian population of the country, production is always short of demand. India has to import pulses to meet the demand for domestic consumption. In the year 2015–16, India imported 5.79 million tons of pulses. Due to high demand for pulses for consumption, not much is left for export, and Indian pulse exports in the year 2015–16 were merely 0.25 million tons, leaving 21.89 million tons of pulses for domestic consumption (Table 9.3).

Gram, pigeon peas, black grams, and green grams, constitute more than 70% of the pulses produced in India. Chickpeas constitute the single largest share of India's pulse exports, amounting to 84.87% of the total pulse

TABLE 9.3 India's Export and Import of Major Pulses

S. No.	Pulses	Share in Total Pulse Exports (2015–16) (%)	Share in Total Pulse Imports (2015–16) (%)
1.	Peas	2.52	38.72
2.	Chickpeas	84.87	17.79
3.	Green grams	2.5	10.03
4.	Lentils	4.60	21.73
5.	Pigeon peas	1.57	7.98

Source: Department of Agriculture, Cooperation and Farmer's Welfare, 2015–16.

exports during 2015–16, mainly due to the kabuli type. On the import side, peas are the most important pulses, amounting to 38.72% of the total pulse imports, followed by lentils, which contribute 21.73% of India's total pulse imports.

9.4.1 Trade Policy on Pulses

The Advance Authorization Scheme (AAS) allows import of pulses for export after domestic processing and value addition. Due to the upsurge in pulse prices, the Central Government prohibited the export of pulses in 2006. With the country's pulse production reaching a record high of 22.4 million tons in 2016–17 (compared to 16.3 million tons in the previous year), in September 2017, the government lifted the ban on the export of chickpea, black gram, and green gram to help farmers earn remunerative prices.

Kabuli chana, organic pulses, and lentils are allowed for export in limited quantities, provided the export contracts are registered with Agricultural and Processed Food Products Export Development Authority (APEDA), exports are made through customs electronic data interchange (EDI) ports, and APEDA certifies the produce as organic. Since June 2014, exports to Bhutan have been exempted from any ban without any quantitative restrictions. The export of pulses has been permitted in restricted quantities to the Republic of Maldives from 2014 to 15 to 2016–17. By contrast, the import of pulses is free of any quantitative restrictions.

On the import front, the government had reduced the allowable stock limit of pulses to 350 metric tons in October 2015, which was later reversed in November 2015. Now, import of pulses is free without any quantitative restrictions. The bound, statutory, and applied duties on some of the important pulses are given in the Table 9.4.

TABLE 9.4 Import Duty on Important Pulses

	Tariff Rates		
Commodity	**Bound Duty (%)**	**Statutory Duty (%)**	**Applied Duty (%)**
Peas	50	50	0
Chickpeas	100	30	0
Green/black gram	100	30	0
Lentils	100	30	0
Pigeon peas	100	30	10

Source: Department of Agriculture, Cooperation and Farmer's Welfare, 2015–16.

9.4.2 SWOT Analysis of Domestic Production Vis-à-Vis Import of Pulses

The demand for pulses in the country can be met through either domestic production, import, or a combination of both. The extent to which domestic production and import are combined has varied implications for the stakeholders like producer farmer, processor, and consumer. The SWOT analysis of domestic production vis-à-vis imports in the context of different stakeholders like producer farmer, processor, and consumer in the context of India is summarized in Table 9.5.

9.4.3 The Dal Shock of 2015–16

Due to a sharp rise in pulse prices in the earlier part of 2016, bullish expectations prompted Indian traders enter into forward contracts which were speculative in nature and made without any knowledge about the crop prospects in India and abroad. After July 2016, the pulse market across the world underwent a dramatic change, with countries including India planting record acreages. This, coupled with a well-distributed southwest monsoon, resulted in a record harvest. The domestic and international pulse prices of pulses plummeted. Following this, there were several contract defaults and forced price negotiations, sending the wrong signals about India to the world pulse market.

9.5. LENTIL PRODUCTION IN INDIA

Lentils are grown as a *rabi* (winter season) crop in India. India ranks first in terms of area and second in production for lentils in the world. Canada ranks first in production of lentils because of the high productivity of 1632kg/ha. Canada, India, Turkey, Australia, Nepal, and Bangladesh are the lentil major producers. In India, lentils are produced in Madhya Pradesh, Uttar Pradesh, Bihar, West Bengal, Rajasthan, and Assam. States like Madhya Pradesh, Uttar Pradesh, and Bihar contribute more than 80% of India's lentil production. The temporal pattern of growth in area, production, and yield of lentils in India is given in Table 9.6.

9.5.1 Economic Classification

Based on the seed size and test weight, lentils are classified into two main groups:

(i) **Bold-seeded**: Includes sub sp. *macrosperma*, with the test weight of more than 25 g. Also known locally as *Masur*or *MalkaMasur,* and mainly cultivated in Bundelkhand region of UP/MP and Maharashtra state.
(ii) **Small-seeded**: Sub sp. *microsperma*, test weight less than 25 g, locally known as *masuri,* and primarily grown in the Indo-Gangetic plains of NEPZ (UP, Bihar, West Bengal, and Assam) (Tiwari and Shivhare, 2016).

TABLE 9.5 SWOT Analysis of Domestic Production Vis-à-Vis Imports

S.·No.	Stake-Holders	Strength	Weakness	Opportunity	Threat
(A)	*Domestic Production*				
	(i) Farmer	Increased income and employment of family labor	Low and unstable yield High cost of production	Enhanced income by attaining yield potentials and taking up processing activities	Uncertainty in production Yield risk Price risk
	(iii) Consumer	Continued choice for locally produced pulses	Has to pay higher price	Choice according to taste	Rising prices
(B)	*Import*				
	(i) Farmer/ processor	Choice for alternative crop	Disincentive for continued cultivation	Export after processing	Shift in cropping pattern toward less remunerative corps
	(iii) Consumer	Availability at low price	Compromise with taste	Price advantage due to reduction in demand supply gap	International price increase

TABLE 9.6 **Temporal Trend and Compound Growth Rate of Lentil Production in India**

Year	Area (Million ha)	Production (Million tons)	Yield (kg/ha)
2005–06	1.51	0.95	629
2006–07	1.47	0.91	621
2007–08	1.31	0.81	622
2008–09	1.38	0.95	693
2009–10	1.48	1.03	697
2010–11	1.60	0.94	591
2011–12	1.56	1.06	678
2012–13	1.42	1.13	797
2013–14	1.34	1.02	761
2014–15	1.47	1.04	705

Source: Agricultural Statistics at a Glance, 2016. Directorate of Economics and Statistics, Department of Agriculture, Cooperation and Farmers Welfare, Ministry of Agriculture, Government of India, New Delhi.

9.5.2 The Lentil Trade Scenario

Due to the mismatch between supply and demand for pulses, the prices of pulse crops have increased exorbitantly. To meet the demand for pulses, India has been importing a large quantity of them in recent years (Reddy and Reddy, 2010). During 2015–16, out of the total 5,797,000,000 tons of pulses imported by India, almost 39% were peas, 22% were lentils, 18% chickpeas, 10% green and black gram, and 8% were pigeon peas. Lentils are the second most imported pulses in India, second only to peas. India imports lentils from Canada, the United States, Australia, Turkey, and Mozambique.

9.6. COMPARATIVE ECONOMICS OF COST OF PRODUCTION OF LENTILS IN MADHYA PRADESH

The details of cost of cultivation for lentils for the years 2007–08 and 2014–15 are given in Table 9.7.

During the period 2007–08 and 2014–15, the operational cost for lentil cultivation has increased by around 170%, and the fixed cost has increased by around 80%. Consequently, the total cost of cultivation has increased by 125%. This increase in variable cost has been mainly due to increasing costs for machine labor,

TABLE 9.7 BreakDown of Cost of Cultivation per Hectare of Lentils in India (2007–08 and 2014–15) (Rs./ha)

Items	2007–08	2014–15
Human labor	2223.56	6127.48
Bullock labor	585.67	1032.69
Machine labor	1277.09	4503.80
Seed	1336.53	2938.64
Fertilizer and manure		
Fertilizer	383.97	1309.88
Manure	0.00	0.00
Total	383.97	1309.88
Insecticides	176.09	311.27
Irrigation charges	697.92	1855.89
Int. on working capital	165.35	442.90
Miscellaneous	6.46	8.53
Fixed cost	6966.00	12,590.35
Rental value of owned land	5864.08	9388.68
Rent paid for leased land	0.00	0.00
Land revenue, ceases, and taxes	5.89	5.84
Dep. on implements and F. Flug.	280.58	500.11
Int. on fixed capital	815.45	2695.72
Total cost	13,818.64	31,121.43

Source: CACP Reports by Ministry of Agriculture and Farmers Welfare and GoI, New Delhi (2011 and 2016).

fertilizers, human labor, and seeds. The costs of the first two have increased by more than 200%, and the latter two, by more than 100%. The increase in machine labor could be a result of a hike in diesel prices during the period. Similarly, the irrigation charges have gone up by about 165% during this period, which again is largely due to an increase in electricity charges and diesel prices. The total cost of cultivation has increased by around 125% during this period (Table 9.8).

The data in the above table makes it clear that the cost of cultivating lentils has increased over the years from 2008 to 09 to 2014–15. While the sum of cost A2 and family labor have increased by around 60% during this period, the cost

TABLE 9.8 Economics of Lentils in India

Year	Cost of Cultivation (Rs./ha)		Net Returns (Rs./ha)
	Cost A2 + FL	Cost C2	
2008–09 to 2010–11	10,057	18,212	8551
2009–10 to 2011–12	10,896	19,076	8049
2010–11 to 2012–13	12,114	20,878	8399
2011–12 to 2013–14	13,941	23,343	8317
TE 2014–15	16,170	26,653	7557

Source: Commission of Agricultural Cost and Prices, Ministry of Agriculture and Farmers Welfare, Government of India, 2016.

C2 has increased by 46%. The net returns, on the other hand, have decreased over the years. From 2008 to 09 to 2014–15, net returns decreased by around 11%. The main reasons for this decrease might be one or two bad agricultural years. In spite of the surge in pulse prices over the years, it is worth pointing out that the domestic prices of lentils have not been commensurate with the increase in cost of cultivation during this period.

9.7. COMPARATIVE ECONOMICS OF LENTILS AND COMPETING CROPS

The comparative economics of lentils and their competing crops in terms of cost of cultivation, gross return, and profit margin on unit area basis in India are given in Table 9.9.

Pulses in India are grown as subsistence crops on minimum or no irrigation. This is why rabi crops such as wheat that are cultivated on irrigated land with assured irrigation have not been considered competing crops for lentils. But the water requirements of rapeseed, mustard, and barley are much lower than that of wheat, so those crops could be considered to compete with lentils for area. The returns over cost of cultivation are highest for rapeseed and mustard, followed by barley. Lentils, with a net return of Rs. 7557/ha, have the lowest returns over cost. These low returns over cost are the biggest impediment to lentil cultivation in India, whereas crops like barley, which were not in demand few years ago, have gained momentum because of an assured market (mainly the malt industry, in the case of barley).

9.7.1 Prices of Lentils in India

It is evident from the table that the domestic price of lentils has increased at a faster rate than the MSP. A comparison of domestic prices with international

TABLE 9.9 Comparative Economics of Lentils and Competing Crops in India (2014–15)

Crop	Cost of Cultivation (Rs./ha)		Gross Value of Output (Rs./ha)	Return Over Cost (Rs./ha)		Net Returns (Rs./ha)
	Cost A2	Cost C2		Cost A2	Cost A2 + FL	
Lentils	12,728	26,653	34,210	21,482	18,040	7557
Barley	18,898	43,867	53,657	34,759	23,704	9791
Rapeseed and mustard	14,813	35,381	45,421	30,607	22,848	10,040

Source: Commission of Agricultural Cost and Prices, Ministry of Agriculture and Farmers Welfare, Government of India, 2016.

prices shows that during 2009–10 and 2010–11, domestic prices were lower. However, from 2011 to 12 onward, domestic prices have been higher than international prices. This clearly shows that, at existing costs and returns for domestically grown lentils, India lacks a comparative advantage in lentil cultivation; this is the reason India is a net importer of lentils. Lentils' cost of cultivation has increased considerably over the years. However, due to the lack of a major technological breakthrough resulting in new high-yielding and drought-resistant varieties, the net returns from lentil cultivation have remained very low. These cost and return scenarios do not favor an acreage increase for lentils. And the major reason is the sensitivity of this crop, as it cannot withstand high wind, rain, or the attack of major diseases (like Stemphylium blight during blossom) compared to other competing crops, which are hardier (Table 9.10).

9.7.2 Opportunities in Rice Fallows

Rice-wheat is the major crop rotation pattern for food self-security in India. Rice cultivated all over the country occupies an area of about 44 million hectares (Agricultural Statistics, 2014) and is grown under both irrigated and rain-fed conditions in various cropping systems. In irrigated areas, rice-wheat, rice-rice, rice-sugar cane, rice-ground nut, rice-vegetables, and rice-mustard are important crop rotations, whereas, in rain-fed areas, rice-pulses, rice-sunflower, rice-sesame, and rice-fallow are prevalent. Eastern India, the rice-dominated area of the country, accounts for about 63.3% of India's total rice

TABLE 9.10 Comparison of Cost of Production and Domestic Prices of Lentils

Year	Minimum Support Price (Rs./qtl.)	Domestic Price (Rs./qtl.)	International Price (Rs./qtl.)
2009–10	1870	4150.25	4183.25
2010–11	1870	3345.25	3747
2011–12	2250	3267.25	3192.25
2012–13	2800	3733	3499.25
2013–14	2900	4275.75	4140.75
2014–15	2950	5258.75	5120
2015–16	3075	6746.25	5764.25

Source: Commission of Agricultural Cost and Prices, Ministry of Agriculture and Farmers Welfare, Government of India, 2016.

area. About 78.8% of rice area in the region is rain-fed, with the result that rice is grown only during the rainy season (June-September), after which, the land lies fallow, hence the name, "rice fallow."

The existing rice fallow, an area of 11.65 million ha (Subbarao et al., 2001; NAAS, 2013) after the rice harvest is almost equivalent to the net sown area of Punjab, Haryana, and Western Uttar Pradesh — the sheet of green revolution in India. Of the total rice fallow area, about 82% lies in the eastern region (Assam, Bihar, Chhattisgarh, Jharkhand, Madhya Pradesh, Orissa, and West Bengal). If this area is brought under cultivation, it may usher in another green revolution in the country, benefiting millions of poor, deprived, small landholders (Joshi et al., 2002; Agricultural Statistics, 2011). The introduction of short-duration pulses like lentils, grass peas, and chickpeas, (which should fit the cropping sequence) will not only provide nutritional security, improve soil health, and increase cropping intensity, but also will increase income. Interventions by the ICAR and CGIAR institutes revealed that lentil varieties like HUL-57, Malika, and Subrata performed well, yielding up to 1.1 t/ha in rice fallows. With proper technology targeting and upscaling, there are about 1–2 million ha in rice fallows areas which could be potentially planted in lentils. This will not only help increase lentil area and production in India, but will also provide an extra source of income for farmers.

9.8. CONCLUSIONS

India ranks first in area, second in production, and lags behind several countries in lentil productivity. In India, lentils are mainly produced in the BIMARU (Bihar, Madhya Pradesh, Rajasthan, and Uttar Pradesh) states as a rain-fed crop. India is a net importer of pulses, and lentils are the second most-imported pulse, second only to peas. Over the past decade, lentils' production cost has more than doubled. The net returns for lentils have shown a declining trend, as there has been no substantial breakthrough in yield-enhancing varieties. Increasing cost of cultivation, coupled with low yields, renders this crop less profitable than other competing rabi crops. Additionally, lentils' low Minimum Support Price, compared to domestic and international prices, discourages area increase. Higher domestic prices, as compared to international prices, encourage more imports. These price dynamics are not favorable when it comes to enhancing domestic lentil production. Enhancing domestic lentil production requires a two-pronged approach: enable research favoring development of high yielding draught resistant varieties, and set price policy to favor acreage increase.

REFERENCES

Agricultural Statistics at a Glance, 2011. Directorate of Economics and Statistics, Department of Agriculture, Cooperation & Farmers Welfare, Ministry of Agriculture, Government of India, New Delhi.

Agricultural Statistics at a Glance, 2014. Directorate of Economics and Statistics, Department of Agriculture, Cooperation & Farmers Welfare, Ministry of Agriculture, Government of India, New Delhi.

Ali, M., Gupta, S., 2012. Carrying capacity of Indian agriculture: pulse crops. Curr. Sci. 102, 874–881.

Al-Issa, Y., 2006. Commodity Brief No. 7 Lentils and Chickpea. National Agricultural Policy Center, FAO, Rome, Italy.

FAO, 2016. Food Outlook: Biannual Report on Global Food Markets. 139. FAO.

Harlan, J.R., 1992. Crops and Man. American Society of Agronomy, Crop Science Society, Madison, WI, USA.

International Institute for Population Sciences (IIPS) and Macro International, 2007. National Family Health Survey (NFHS-3), 2005–06: India. vol. I. Demographic and Health Surveys, Mumbai, India.

Joshi, P.K., Birthal, P.S., Bourai, V.A., 2002. Socioeconomic Constraints and Opportunities in Rainfed Rabi Cropping in Rice Fallows Areas of India. National Centre for Agricultural Economics and Policy Research, New Delhi.

Kerr, R.B., Snapp, S., Chirwa, M., Shumba, L., Msachi, R., 2007. Participatory research on legume diversification with Malawian small holder farmers for improving human nutrition and soil fertility. Exp. Agric. 43, 437–453.

McDermott, J., Wyatt, A.J., 2017. The role of pulses in sustainable and healthy food systems. Ann. N. Y. Acad. Sci. Issue: Potent. Pulses Meet Today's Health Challenges: Staple Foods.

NAAS, 2013. Improving productivity of rice fallows, policy paper 64. National Academy of Agricultural Sciences, New Delhi.

Reddy, A.A., 2004. Consumption pattern, trade and production potential of pulses. Econ. Polit. Week. 4854–4860.

Reddy, A.A., Reddy, G.P., 2010. Supply side constrains in production of pulses in India: a case study of lentil. Agric. Econ. Res. Rev. 23, 12.

Roy, D., Kishore, A., Joshi, P.K., Mishra, B., 2016. Pulses for food and nutritional security: the technology perspective. Indian J. Genet. Plant Breed. 76 (4), 375–387.

Sathe, D., Agarwal, S., 2004. Liberalisation of pulses sector production, prices and imports. Econ. Political Week. 3391–3397.

Sell, R., 1993. Lentil. In: Alternative Agriculture Series, No. 15, July 1993. North Dakota State University, USA.

Sharma, B., 2009. In: Erskine, W., Muehlbauer, F.J., Sarker, A., Sharma, B. (Eds.), Genetics of economic traits. The Lentil Botany Production and Uses, CABI International, UK.

Singh, K.M., Singh, A.K., 2014. Lentil in India: An Overview. SSRN Electr. J. https://doi.org/10.13140/2.1.2191.1049.

Subbarao, G., Kumar Rao, J., Kumar, C., Johansen, U., Irshad, A., Krishna Rao, L., Venkataratnam, K., Hebbar, K., Sai, M., Harries, D., 2001. Spatial Distribution and Quantification of Rice-fallows in South Asia: Potential for Legumes. ICRISAT, Hyderabad, pp. 316.

Tiwari, A.K., Shivhare, A.K., 2016. Pulses in India: Retrospect and Prospects. Govt. of India, Ministry of Agri. & Farmers Welfare (DAC&FW), Directorate of Pulses Development, Vindhyachal Bhawan, Bhopal, MP.

FURTHER READING

Agricultural Statistics at a Glance, 2016. Directorate of Economics and Statistics, Department of Agriculture, Cooperation & Farmers Welfare, Ministry of Agriculture, Government of India, New Delhi.

CACP, 2011. Reports of the Commission for Agricultural Costs and Prices for the Crops Sown during 2010–11 Season. Commission for Agricultural Costs and Prices, New Delhi.

CACP, 2016. Price Policy for Rabi Crops: The Marketing Season 2017–18. Commission for Agricultural Costs and Prices, New Delhi.

Gopalan, C., Rama Sastri, B.V., Balasubramanyan, S.C., Narasinga Rao, B.S., Deosthale, Y.G., Pant, K.C., 2004. Nutritive Value of Indian Foods. National Institute of Nutrition, ICMR, Hyderabad.

Index

Note: Page numbers followed by *f* indicate figures and *t* indicate tables.

A

Abiotic stress
  biotic and, 1–4
  lentil, 142–144
Advance Authorization Scheme (AAS), 210
Aethiopicae, 10–11, 13–14
Agricultural and Processed Food Products Export Development Authority (APEDA), 210
Agrobacterium-mediated genetic transformation, 3–4, 146, 153
  fungal disease resistance genes, 149–150
  gene construct, 153–155, 154*f*
  GUS histochemical assay, 156
  marker genes, 175–188
  molecular analysis in transgene expression, 156–157
  selection of transformants, 155–156
  status of, 147–149
  steps of, 155–156
  transgenic progenies, analysis of, 157–158
Agrobacterium rhizogens, 148–149
Agrobacterium tumefaciens, 147–149, 153–158
Amplified fragment length polymorphism (AFLP) markers, 99
Aneuploids, 59, 63–64
Animal source foods (ASF), 203–204
Antifungal protein *(AFP)* genes, 149–150, 155
  amplification, 157, 194*f*
  transformation with LBA4404/pCAMBIA2300 enh35S, 191–195
Antifungal transgenic approaches, 149–150
Arabidopsis thaliana, 155
Ascochyta blight, 84–85, 134–135, 142–144
Association mapping (AM), 127–128, 131–134

B

Bangladesh Agricultural Research Institute (BARI), 151
Benzyl amino purine (BAP), 147
Biotechnology application, 145–150
Biotic stress, 142–144
Biparental mapping populations, 127–128, 131, 132–133*t*
Black-box approach, 30
Bold-seeded lentil, 211
Breeding methods, 43–44
  conventional breeding, 175
    integrating multiple traits, 85–97
    lentil improvement through, 144–145
  genomic estimated breeding value, 114–115, 114*f*
  haploid breeding, 73–74
  independent culling, 89
  marker-assisted, 131, 134
  molecular breeding, 97–115
  for multiple traits, 85–97, 86*f*
  mutation breeding, 57–58, 64
  pre-breeding method, 145–146
  selection index, 97
  selective breeding, 46
  self-fertilizing crops, 51–52
  self-pollination, 51–52
  tandem selection, 89

C

C-banding technique, 62–63
Chickpeas. *See also* Pulses
  production, 204, 207*t*
  transformed plantlets, 177
Chitinase gene, 150, 155
  PCR amplification, 190, 193*f*
  product size, 157
4-Chloroindole-3 acetic acid (4-Cl-IAA), 72
Chloroplast DNA (cpDNA), 51, 67
Chromosome analysis, 46–47
Chromosome banding techniques, 62–63
Colchicine, 61–62
Colchiploidy, 61–62
Comparative economics
  cost of production of lentils, 213–215
  lentils and competing crops, 215–218, 216*t*
Competing crops, economics, 215–218, 216*t*

Conventional breeding, 175
integrating multiple traits, 85–97
lentil improvement through, 144–145
Conventional manipulations, 47–48
Core collections development, 34–35
Cotyledonary node (CN) explant
GUS histochemical assay, 178*f*, 179*t*
in-vitro regeneration, 151, 160
on MS medium, 151, 158–159, 161*t*, 162*f*
multiple shoot buds from, 158–160, 159*f*, 162*f*
Cotyledon-attached decapitated embryo (CADE) explant, 3–4, 191
GUS histochemical assay, 178*f*, 179*t*
in-vitro flower and pod formation, 152, 187*t*
in-vitro regeneration, 151, 160
kanamycin effect, 181*f*, 182*t*
on MS medium, 151, 158–159, 161*t*
Cowpeas, 204, 207*t*. *See also* Pulses
Cropping systems, 203–204
Crop wild relatives (CWRs), 2–3
CSK Himachal Pradesh Agricultural University, 127–128
CTAB method, 156–157
Cytogenetic manipulation, 46–47, 59–61
biochemical and molecular biology approach, 59
chromosome banding techniques, 62–63
classical, 60
molecular cytogenetic approaches, 64–67
mutation studies, 63–64
and ploidy status, 61–62
traditional and molecular, 59

D

Dal Shock of 2015–16, 211
Decapitated embryo (DE) explant
GUS histochemical assay, 178*f*, 179*t*
in-vitro regeneration, 151, 160
on MS medium, 151, 158–159, 161*t*
transformation with LBA4404 strain, 177
DH approach. *See* Double-haploid (DH) approach
DNA-based molecular markers, 101–102*t*
Double-haploid (DH) approach, 73–74
DREB1A gene, 149
Dry beans, 204, 207*t*. *See also* Pulses
Duplicates, 30

E

Economic classification, 211, 215*t*
Embryo-ovule rescue technique, 59, 67–73, 70*f*
Ervoides, 10–11
EURISCO system, 26
Explants. *See also specific explants*
Agrobacterium-mediated genetic transformation, 177–178
multiple shoot formation, 158–160, 159*f*, 161*t*, 162*f*
types of, 151, 153, 158, 176–177
Expressed Sequenced Tags (EST), 126
Ex-situ conservation, 24–26, 25–29*t*
Ex-situ germplasm collections, 2–3

F

Fabaceae, 1–2
Feulgen stain, 46–47, 66
FISH. *See* Fluorescence in situ hybridization (FISH)
Flanking markers, 131, 132–133*t*, 134–135
Fluorescence in situ hybridization (FISH), 3
cytogenetic manipulation, 46–47
molecular cytogenetics, 64–67
Focused Identification of Germplasm Strategy (FIGS), 3–4
Fungal disease, 142–144, 149–150
Fusarium oxysporium, 142–144
Fusarium wilt-resistance (Fw) gene, 131

G

G-banding technique, 62
Gene bank Information Management System (GIMS), 26
Gene pools, 14–16
classification, 58–59, 58*f*
genetic diversity, 134
primary and secondary, 3, 44–45, 48
species distribution and, 22
tertiary and quaternary, 67–68
Gene pyramiding, 3–4, 109
marker-assisted backcrossing, 110–112
convergent/separate backcrossing, 110–112, 112*f*
simultaneous backcrossing, 110–112, 111*f*
stepwise backcrossing, 110–112, 111*f*
through multiple parents crossing, 109–110
Gene stacking, 109
Genetic correlation, 88–89, 90–96*t*
Genetic diversity, 47
Genetic engineering, 46
Genetic linkage maps, 100
Genetic manipulation, 46–51
Genetic transformation, 3–4

Agrobacterium-mediated (*see* *Agrobacterium*-mediated genetic transformation)
antifungal gene constructs, 188–195
in-vitro flowering and pod formation, 165–166, 174*f*
anthers from, 166, 170*f*
auxins concentrations effect on, 168*t*
bitter melon, 167
development, 166–167, 167*f*, 169*f*
explants for, 173*t*
pollen grains, 166–167, 170*f*
protocol for, 152–153, 165*f*
transformed shoots, 184–185, 185–186*f*
in-vitro regeneration, 146–147
explants types, 151
plant materials used for, 151
root development, 151–152, 160–163
shoots development, 151, 158–160
Genome in-situ hybridization (GISH), 3, 64–65
Genomic estimated breeding value (GEBV), 114–115, 114*f*
Genomic selection, 114–115, 114*f*
Genotype×environment (G×E) interactions, 125–126
Genotyping by Sequencing (GBS) method, 3, 15–16
Germplasm, 30
characterization and evaluation, 30–32
conservation, 22–30
quality traits, 31*t*
Giemsa C-banding technique, 62
GISH. *See* Genome in-situ hybridization (GISH)
Global Crop Diversity Trust (GCDT), 24
Grain legumes. *See* Pulses
GUS reporter gene, 147–148, 155
chimeric expression, 183–184, 184*f*
histochemical assay, 156, 177, 183
PCR amplification, 148, 157, 188, 188*f*
putative transformed shoots, 183, 183*f*

H

Haploid breeding, 73–74

I

Illumina GA/GAIIx system, 126
Illumina Genome Analyzer, 126–127
Illumina GoldenGate (GG), 126–127
Independent culling, 89–97
Index selection, 97, 115
India
lentil in, 211
cost of cultivation, 213–215, 214*t*
economic classification, 211, 215*t*
prices of, 215–217, 217*t*
temporal trend and compound growth rate, 213*t*
trade scenario, 213
pulses in, 209–211, 209*t*
Indian Institute of Pulses Research (IIPR), 127–128
Indole-3-butyric acid (IBA), in-vitro shoots
flowers, 166, 167*f*
in-vitro pods formation, 172*f*
root induction, 164–165*f*, 164*t*
Insect pests, lentil, 142–144
In-situ conservation, 23–24
Intermediae, 10–11
International Centre for Agricultural Research in Dry Areas (ICARDA), 2–3, 22–24, 25*t*, 26, 27–29*t*
Intersimple sequence repeat (ISSR) markers, 99, 128–131
Interspecific hybridization, 51, 64–65, 68–69, 70*f*
Intraspecific linkage map, 128–131
Introgression, 1–2, 46–48, 59, 67–68
In-vitro flowering, 165–166, 174*f*
anthers from, 166, 170*f*
auxins concentrations effect on, 168*t*
bitter melon, 167
development, 166–167, 167*f*, 169*f*
explants for, 173*t*
pollen grains, 166–167, 170*f*
protocol for, 152–153, 165*f*
transformed shoots, 184–185, 185–186*f*
In-vitro regeneration, 146–147
explants types, 151
plant materials used for, 151
root development, 151–152, 160–163
shoots development, 151, 158–160
In-vitro seed germination, 172–175, 175*f*
In vivo intergeneric grafting techniques, 72

K

Kanamycin, 155–156
albino shoots development, 180, 180*f*
chickpea, 181–183
effect on developing shoots, 180–181, 181*f*, 182*t*
GUS expression, 183–184
in-vitro flower on shoot recovered through, 185*f*
*npt*II gene, 191

L

Legumes. *See* Pulses
Lentil (*Lens culinaris* Medik.), 1–2, 7–8, 43–44, 57–58, 83, 125, 206–209
  BAP effect on root formation, 147
  biotechnological approaches, 145–150
  biotic and abiotic stress factors, 142–144
  carbohydrates, 125
  comparative economics of, 215–218, 216*t*
  core collections development, 34–35
  cultivated in countries, 141–142
  cytogenetic manipulation, 46–47
  distribution, 13–14
  domestication, 16–17
  duplicates, 30
  ex-situ conservation, 24–26
  floral biology, 11–13
  fungal diseases, 142–144
  gene distribution, 45*f*
  gene pool, 14–16
  genetic linkage maps of, 100
  genetic manipulation, 46–51
  germplasm characterization and evaluation, 30–32
  germplasm conservation, 22–30
  history, 8–9
  improvement through conventional breeding, 144–145
  in India, 211
    cost of cultivation, 213–215, 214*t*
    economic classification, 211, 215*t*
    prices of, 215–217, 217*t*
    temporal trend and compound growth rate, 213*t*
    trade scenario, 213
  insect pests causing damage to, 142–144
  in-situ conservation, 23–24
  morphology, 11–13
  morphophysiological variation, 50*t*
  multiple shoot formation, 146–147, 158–160, 161*t*
  origin, 9–11
  protein content, 141
  region-wise targeted traits, 87*t*
  safety duplicate, 30
  salinity affect, 144
  seeds, 142, 144*f*
  species distribution and gene pools, 22
  taxonomy, 11–13, 44–46, 58–59
  trait of interest, 33–34*t*
  wild species, 44–47, 49–50, 49*t*
Linkage map development, 99–100

M

Macrosperma, 10–11, 43–44, 46–48, 142
Marker-assisted backcrossing (MABC), 103–109, 108*f*, 134–135
Marker-assisted backcrossing-gene pyramiding, 110–112
  convergent/separate backcrossing, 110–112, 112*f*
  simultaneous backcrossing, 110–112, 111*f*
  stepwise backcrossing, 110–112, 111*f*
Marker-assisted breeding program, 131, 134
Marker-assisted recurrent selection (MARS), 112–113, 113*f*
Marker-assisted selection (MAS), 126
  backcrossing program, 134–135
  breeding materials, 134
  gene pyramiding, 98, 98*f*, 100, 104–107*t*
Marker genes, *Agrobacterium*-mediated genetic transformation
  genetic transformation, 175–178, 179*t*
  in-vitro flowering and pod formation, 184–185, 185–186*f*
  regeneration/selection of transformed shoots, 178–184, 180*f*
  transgene integration, analysis of, 188
Marker-trait association analysis, 131–134
Marker-trait linkage analysis, 125–126, 131, 132–133*t*
Mature seeds, 153
Micrografting technique, 147, 163
Microsperma, 10–11, 43–44, 46, 142
  genetic base, 47–48
  healthy lentil plants growing in field, 142–144, 143*f*
  hybridization, 47–48
  in-vitro regeneration, 147, 150
  mature dry seeds of, 144*f*
  varieties, 151
Molecular analysis, 156–157
Molecular breeding, 97–115, 145
Molecular cytogenetics, 59, 64–67
Molecular linkage maps, 128–131, 129–130*t*
Molecular markers, 35, 84–85, 97–115, 126–127
  deployment of, 125–126
  genetic diversity, 134
  identification, 99–100
Morphological and isozyme markers, 99
Morphophysiological variation, 50*t*
Multiple parents crossing, 109–110
Multiple traits
  breeding schemes for, 89–97
  conventional breeding integrating, 85–97

genetic correlation, 88–89, 90–96*t*
genomic selection, 114–115, 114*f*
for potential genetic improvement, 88*f*
selection, 3–4
Murashige and Skoog (MS) media
embryo-ovule rescue technique, 69, 71–72
explants, 158–160, 162*f*
indole-3-butyric acid, 151–152
with thiadiazuron, 146–147
Mutation breeding, 57–58, 64
Mutation studies, cytogenetic techniques, 63–64

N

National Bureau of Plant Genetic Resources (NBPGR), 26, 26*t*
NATP projects, 26
Next-generation sequencing, 126–127
Northern blot, 157, 193–195
*npt*II genes
coding sequence, 156–157
PCR amplification, 148, 157, 188, 189*f*, 191, 193–195, 194*f*
Nucleolar organizing region-banding (N-banding), 62

O

Optical density (OD), 155–156, 177
Oxyquinoline, 59–60

P

PCR-based markers, 100
PGRdup, 30
Phenotypic selection (PS), 112
Phenotypic Variance Explained (PVE), 132–133*t*
Phosphinothricin (PPT), 156, 188–190, 190–191*f*, 192*t*
Phytohormones, 167
Pigeon peas, 204, 207*t*. *See also* Pulses
Pilosae, 10–11
Plant breeders, 145
Plant genetic transformation, 145–146
experiments on, 191
fungal disease resistance, 149–150
Pod formation, 165–166, 174*f*
anthers from, 166, 170*f*
auxins concentrations effect on, 168*t*
bitter melon, 167
development, 166–167, 167*f*, 169*f*
explants for, 173*t*
pollen grains, 166–167, 170*f*
protocol for, 152–153, 165*f*
transformed shoots, 184–185, 185–186*f*
Pollen grains, 152, 166–167, 170*f*
Polymerase chain reaction (PCR)
chitinase gene, 190, 193*f*
GUS reporter gene, 148, 157, 188, 188*f*
marker genes, 185
*npt*II genes, 148, 157, 188, 189*f*, 191, 193–195, 194*f*
Polyploidy, 61–62
Population-specific linkage maps, 128–131
Pre-breeding method, 145–146
Pulses, 141, 203–204. *See also* Lentil (*Lens culinaris* Medik.)
consumption, 204
Dal Shock of 2015–16, 211
import duty on, 210*t*
in India, 209–211, 209*t*
nutritive value, 205*t*
production, 204–206, 206*f*, 207*t*, 208*f*
SWOT analysis of domestic production vis-à-vis import, 211, 212*t*
trade policy on, 210

Q

Q-banding, 62
Quantitative trait loci (QTL)
Ascochyta blight, 134–135
biparental mapping population, 131–134, 132–133t
gene pyramiding, 84–85, 100–103, 112–113
mapping, 99–100

R

Randomly amplified polymorphic DNA (RAPD) markers, 99
Recombinant inbred line (RIL) population, 127–131, 127*t*, 134–135
Resistance gene analogue (RGA) markers, 99
Restriction fragment length polymorphism (RFLP) markers, 99
Reverse transcription PCR (RT-PCR), 157, 191
AFP gene, 194*f*
*npt*II gene, 194*f*
Rhizactonia solani, 142–144
Rice fallows, 217–218
Rice-lentil system, 8
Root development, in-vitro regeneration, 151–152, 160–163

S

Safety duplicate, 30
Salinity, 144
Salt tolerance, 8
Satellite (SAT) chromosomes, 62–63
Second-generation technology, 126
Selective breeding, 46
Self-pollination
  breeding programs, 51–52
  ovary, 166–167, 171*f*
  pollen viability and tube growth, 152–153
Shoots development, in-vitro regeneration, 151, 158–160
Single Nucleotide Polymorphism (SNP) markers, 126–131
Single-seed descent (SSD), 71–73
Small-seeded lentil, 211
SNP-based linkage map, 100
Southern blot, 149, 157, 193–195
Species distribution, 22
SSR markers, 99, 126–127
Stemphylium blight, 134–135, 142–144, 143*f*
Subspontanea, 10
Sulfate polyacrylamide gel electrophoresis (SDS-PAGE), 67
SWOT analysis, 211, 212*t*

T

Tandem selection, 89
Taxonomy, lentil, 44–46, 58–59, 58*f*
Tissue culture technique, 73–74, 158
Trade policy, on pulses, 210
Trait-specific mapping populations, 127–128
Transcriptome sequencing, 126
Transformed shoots
  *npt*II gene PCR amplification, 189*f*
  phosphinothricin, 189–190
  putative, 183*f*, 185, 188, 190–191, 193–195
  regeneration/selection, 178–184, 180*f*, 182*t*
Transgenic plant technology, 71–72
Transgenic progenies, 157–158
TRIZOL reagent, 157
Truncation selection, 89–97
Tyrosine, 160

U

Uromyces fabae, 142–144
USDA-ARS, 25*t*, 26

V

Vacuum infiltration technique, 148
Vicia, 10–11

W

Waterlogging, 8
Whole genome sequencing, 46
Wild species
  genetic manipulations, 44–47, 49–50, 49*t*
  germplasm collection, 2–3

CPI Antony Rowe
Chippenham, UK
2018-10-11 20:11